The House on Ipswich Marsh

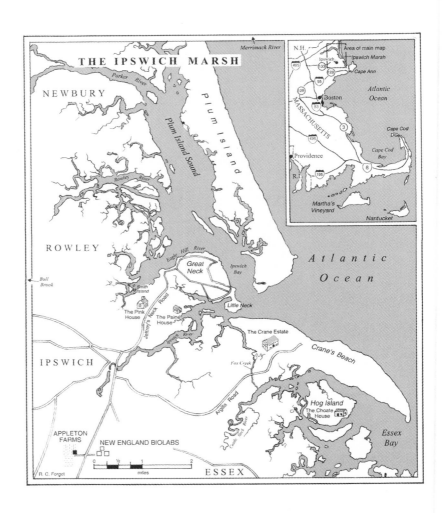

THE IPSWICH MARSH

NEWBURY

ROWLEY

IPSWICH

ESSEX

Merrimack River

Parker River

Plum Island Sound

Plum Island

Rowley River

Bull Brook

Eagle Hill River

Great Neck

Ipswich Bay

Smith Island

Jeffrey's Neck Road

The Pink House

The Paine House

Little Neck

Ipswich River

The Crane Estate

Crane's Beach

Fox Creek

Argilla Road

Castle Neck River

Hog Island

The Choate House

Essex Bay

APPLETON FARMS

NEW ENGLAND BIOLABS

Atlantic

Ocean

R. C. Forget

0   ½   1   2
                miles

N.H.   Area of main map
              Ipswich Marsh
495   Ipswich
128   Cape Ann
95
                    Atlantic
                    Ocean
128
MASSACHUSETTS
93   Boston
                         Cape Cod
3
495   Cape Cod
                        Bay
Providence   6
R.I.   195
                    Martha's
                    Vineyard
                              Nantucket

The House on
Ipswich Marsh

Exploring the Natural History
of New England

WILLIAM SARGENT

University Press of New England

Hanover and London

University Press of New England

www.upne.com

© 2005 University Press of New England

All rights reserved

First paperback edition 2014

Paperback ISBN 978-1-61168-771-2

Manufactured in the United States of America

Designed by Katherine B. Kimball

Typeset in Galliard by Integrated Publishing Solutions

For permission to reproduce any of the material in this book, contact
Permissions, University Press of New England, One Court Street,
Suite 250, Lebanon NH 03766; or visit www.upne.com

The Library of Congress has cataloged the hardcover edition as follows:

Sargent, William, 1946–

The house on Ipswich Marsh : exploring the natural history of New England /
William Sargent.

   p.   cm.

Includes index.

ISBN 1–58465–465–1 (cloth : alk. paper)

1. Natural history—Massachusetts—North Shore.   I. Title.

QH105.M4S26 2005

508.744′s—dc22   2004022118

5 4 3 2 1

To Kristina and Chappell,

who urged me to move to Ipswich

and made it all happen.

Contents

Acknowledgments

Many people have helped me write this book. John Phillips first introduced me to the charms of the North Shore when we were callow young undergraduates. I still remember our crashing a frighteningly sophisticated New Year's Eve party at the Crane estate. It soon became obvious we lacked the requisite bloodlines and had not just flown in from some fashionable European resort. But what I remembered most about Ipswich were the houses resting on drumlins surrounded by open marsh.

Ed Dick sold us our pink house, but we quickly learned that one is not actually the owner of an old house, as much as the temporary caretaker for the next generation. Chris Sammartino and Anne Fitzgerald were the former caretaker artists who planted such beguiling gardens outside and created such an eclectic mix of art and fantasy inside. The Peddricks, who are now our good neighbors across the street, provided stories about when Stacey and Chris grew up in the house and filled its halls with music when they were not on tour with the "Fools." Slews of other caretakers inhabited the house, back to the original Captain John Smith in 1740.

I would also like to thank the voters of the town of Ipswich for their foresight in purchasing the Wendell Estate, whose magnificent fields and marshes will remain forever open. I enjoyed my intellectual tussles with the Conservation Commission over their plans to build a parking lot in the prime corner of the field. I have left some of my more curmudgeonly observations in the book to provide local flavor.

The New England Biolabs Foundation provided me with a small grant to study the behavior of bobolinks in the neighboring fields. Don Comb and Richard Grandoni first introduced me to the company, which will be such a welcome addition to the area. Fred Winthrop, Franz Ingelfinger, Rob Eblinger, and Erik Johnson provided invaluable assistance in my chapters about Crane's Beach, the deer hunt, Hog Island, piping plovers, and the sweet bay magnolias of Ravenswood Park. In writing

Crane Beach? "Crane's Beach"!

about the Trustees' properties I have given popular names precedence over strict orthodoxy. Besides, I find "Crane Beach" virtually impossible to pronounce, if not grammatically suspect, so use the more familiar "Crane's Beach" or simply "Crane's" instead. I use Hog Island over Choate Island purely out of historic and aesthetic preference. I hope the Trustees will not revoke my beach permit for such transgressions.

I thank Charlie Shurcliff for his vast knowledge of all things to do with Crane's Beach and Argilla Road and deigning, on occasion, to visit the "other neck." I thank Bill Wasserman for his goodwill, his politics, and his research on Lyme disease.

Phil Kent introduced me to the clam flats, and the Ipswich Shellfish Company invited me to visit its immaculate shucking facility. Lisa Manzi, Kerry Mackin, Jim Berry, Bob Gravino, and Jim Graffum provided help on vernal pools, the Ipswich River, and the Wendell Estate. Bob Oldale and Jim Skehan provided assistance on geology, and Bernd Heinrich and the intrepid students in his winter ecology course at the University of Vermont collected much of the data used in my winter chapters. If I have made any mistakes it is because I am a misplaced Cape Codder still getting used to the cooler waters of the North Shore.

Finally, I would like to thank Kristina and Chappell for sharing our house, our life, and our adventures amidst the marshes of Ipswich Bay.

Introduction

The Pink House

(April 23, 2004)

I fell in love with a field in the spring of 2001. I hadn't expected to fall in love with the field. I had moved to Ipswich to write a book about marine biology. But the field beckoned and there is no accounting for whom one falls in love with, nor how it will affect your life.

It was a long, broad field surrounded by marshes that led to a large, red, open-doored barn that sat on a hill dotted with pear and apple blossoms. Orchard orioles and bobolinks flitted from tree to tree. I trespassed egregiously to explore the farm, even making not-so-subtle inquiries as to the availability of the house and barn. It turned out the house was for sale but at a price way beyond what I could afford.

A year later my wife opened the *Boston Globe* to the real estate page. There was a large pink house sitting beside a swale of phragmites. It was about the most aesthetically and environmentally incorrect an image as could possibly be conceived: a colonial house painted pink, surrounded by one of the most invasive species known to mankind. But there it was: large, pink, and utterly charming.

A real estate agent drove us up to visit the house and I couldn't believe my eyes. The house sat in the corner of the same field I had discovered the year before. Up close it was even better. A carpet of the most brilliant red poppies nodded their heads by the front door; foxglove and hollyhocks swayed in the English-style cottage garden out back. Wisteria draped from the eves and sparrows darted in and out of pink Victorian birdhouses above the portal. A thousand rosebuds bobbed above a white picket fence that wrapped halfway around the front and corner of the house.

A neat, black and white sign announced that this was the house of Captain John Smith built in 1740. Beside the sign sat the voluptuous

"There's just no accounting for whom one falls in love with." The Field.

The old red barn.

The large pink house surrounded by phragmites.

pink torso of a nude mermaid, whose tail worked admirably as a door-knocker. Inside, the twin themes continued. The kitchen boasted modern, pale white birch cabinets overlooking a blue tile floor that made it look like you were stepping into a Mediterranean swimming pool. The dining room felt like the inner cloister of an Italian palazzo, and each bathroom looked like a set from the Little Mermaid.

The agent tried to show us several other houses but we would have none of it. How could we not buy this charming monstrosity! Later we would discover the house was the farmhouse for the original farm that owned the ninety-acre field, and the owners used to churn butter in the basement dairy and make a good living from mowing the marshes for hay.

But the day we moved in, the town announced it was going to build a parking lot beside our back door. I swung into action, writing letters, protesting, and making a general nuisance of myself. In the process I met a lot of nice eccentric new neighbors and was told that it was quite all right to make a ruckus and in fact I wouldn't really be accepted as a true Ipswichite until I did. I also discovered something about dueling traditions. When I mentioned to a neighbor that the local historical society might make me change the color of the house to a more suitable colonial

The Captain John Smith House.

Open fields that stretch to the nearby marsh.

color, she was horrified: "No you can't possibly do that, everyone gives driving directions by your house, 'turn left at the big pink house.' Why, it would be positively unthinkable!"

Eventually I lost the argument about the parking lot, but like to believe I won on points. Now I welcome with open arms the many birders and naturalists who come to park in "our" lot and walk on the 200 acres of open fields that stretch from Ipswich River to the marshes of Eagle Hill.

This then is my story of a field, a marsh, a house, and about a billion years of the biological and geological history of this little corner of the planet known as the North Shore of Boston.

Part I

SPRING

PRECEDING PAGE: *A blanket of fog nestles in the swales of a distant pasture.*

Chapter 1

Spring Dawn

An Awakening

(May 4, 2003)

It is 5:30. A cardinal sings from the limb of a cherry tree and sparrows twitter in the nearby wisteria. The sun has yet to rise and I have no intention of getting up. But nature calls. No, not the ethereal sort, not the poetic call of nature that should come with the first day of spring, but the simple call of nature that comes from a distended bladder. Whatever the reason, the result is the same. I'm up and the fields are covered with a thin layer of frost and a blanket of fog nestles in the valleys and swales of the distant pasture. I grab my cameras and head outside.

There is no wind — no sounds save for the plaintive call of a killdeer in the distance. I make my way toward a shallow slough at the edge of our pasture. My footsteps leave a dark trail through the fragile whiteness of frost. Filaments of mist waft through blades of canary grass emerging from the wetlands. Each new green shoot casts an exact reflection of itself in the quiet shallow waters. A flock of glossy ibis probe the soft mud with their gracefully curving bills and a greater yellowlegs bobs and weaves on the far shore. The slough reflects the erect head of a Canada goose standing sentinel beside his mate.

I feel like I am walking beside a tundra pool high above the Arctic Circle. Frost edges the creeks that meander through the nearby marsh and scudding blankets of fog obliterate the outside world. But the call of a distant pheasant breaks the illusion. I am walking through an extraordinary plot of meadows, marshes, pastures, and islands only forty miles north of Boston.

But it has been an unusually cold and long winter. Snow has lain on the ground from Thanksgiving to Easter. Rabbits and deer have gnawed the bark off the Russian olives invading an old apple orchard. By December, the ice was so thick that neighbors dragged their fishing shacks out

Fishing shacks on the Ipswich River.

of ancient barns and planted them on the Ipswich River. It was the first time in twenty years that the ice had been thick enough to support the single room fishing houses.

In late March a harrier had started to sit in the black locust beside our house. The large hawk was obviously hungry and emaciated. The snow had been so deep that he had been unable to catch the mice and voles that make up the bulk of his wintertime diet. We watched him as he sat hunched over, clinging to the same branch day after day as the snow swirled about his head. He moved as little as possible to conserve energy. We wondered if he would survive the winter.

Beneath him, meadow voles were scurrying happily through their warm humid tunnels deep beneath the snow. The tunnels were well stocked with caches of seeds and grasses the voles had harvested the summer before. While the voles were enjoying the pleasures of their sub-nivean world, snow was swirling over their nemesis above.

But by early April the tables had finally turned. I discovered feather marks in the snow where the harrier had swooped down on a vole who had not realized that the snow had become too thin to mask his quiet squeaks. Even in May a six-inch bank of ice still lurked in the phragmites-filled swale beside our house.

A night heron.

Yet with the warmth of this rising sun we seem to pass from winter to spring. It takes only fifteen minutes for the sun to melt the frost-covered fields and turn them into bright green spring pastures. Each blade of grass once etched with frost is now bedecked with pendant jewels of sun reflecting dew.

But there is poignancy in these wetlands. I have photographed more than twenty species of shorebirds in this slough. But it may soon be drained to protect a neighbor's driveway. I can't dispute the neighbor's right to protect his driveway, but New England will lose one of its premier spots to watch and photograph waterfowl. Another exquisite little gem of nature will have been nickeled and dimed to death.

Walking the path toward Smith Island is a more pleasing prospect. The path leads the eye toward a graceful curve of linden trees. An immense bank of weeping willows nestles beside some conifers. The brilliant yellows of their flowing catkins stand out in stark contrast to the dark greens of the pines and the soft greens of the spring grass. Boulder-strewn causeways lead to drumlin islands that seem to float above the mist and acre upon acre of greening marshlands.

Can it be that this extraordinarily pleasing vista of islands, hills, marshes, and fields is simply the result of millions of years of biological

A young family of Canada geese.

Glossy ibis.

Smith Island.

and geological coevolution? Did this crustal plate, once a part of Africa, simply become a suspect terrane clumped onto the edge of our continental mass? Did glaciers just sculpt these terraces and leave decorative islands behind? Did plows and cows simply create these fields and pastures? Did the sea level simply rise to make this marsh? Did a farmer just plant these trees to feed his livestock and family? Or was there something else going on? Did someone consciously design this land to mimic the evolutionary landscape inherent in our early genes?

As I raced home, I realized that my book had finally started. I had discovered my theme. Was this land merely the result of random coevolution or was it shaped by the hand of conscious design? Was this a natural phenomenon or a man-made landscape cleverly designed to mimic nature?

Perhaps my search had not started as it should, on the first day of spring with an epiphanous call to nature, but halfway through the season with a call of a far more humble sort. Whatever the reason, I had stepped outside and discovered a quest that could last a lifetime.

Chapter 2

Return of the Bobolinks

(May 4, 2003)

This flashing, tinkling meteor bursts through the expectant meadow air, leaving a train of tinkling notes behind.

— HENRY DAVID THOREAU

I was quietly kvetching to myself when I first heard the ebullient call of the bobolink. It had gone below freezing the night before; New Hampshire's Old Man of the Mountains had just plummeted into Franconia Notch. The town fathers would probably rebuild the Old Man, as some sort of ersatz epoxy Mount Rushmore — would this winter never end?

> Robert of Lincoln, Robert of Lincoln,
> Sings his name, sings his name.
> Bob-o-link,
> Bob-o-link,
> spink, spank, spink.

It was the joyous call of a male bobolink tumbling out of a leafless linden tree high overhead. He had arrived the night before from his wintering grounds on the broad pampas of Argentina. He had started his migration in early March, flitting through the towering teak and mahogany canopies of the Amazonia rain forest. His small flock had poured out of the jungles of Central America and flown nonstop across the Gulf of Mexico. Perhaps they had flown over mountains and island hopped up the Caribbean chain. They had migrated by night and gorged by day. Now the male was sitting in the top of this recently frosted tree waiting for the sun to warm its bark enough to awaken its slumbering insects. Is there no better symbol for spring than the reappearance of these joyous little birds, the return of life to this recently frozen land?

The male's rump and sides are elegantly attired with black and white striped plumage, and he wears a resplendent yellow cape that advertises

Male bobolink in the tree overhead.

the nape of his brilliant neck. He looks vaguely exotic but ultimately comic, much like an inebriated swell desperately trying to maintain his dignity with his tuxedo on inside out.

Another male arrives and the two fly out in a broad loop over the field. Their wings flutter but their forward motion seems hardly enough to keep them aloft. This is their courtship flight, but the females have yet to arrive. Perhaps the males fly more to impress each other than their nonexistent mates. Afterward the two sit side-by-side, curiously amiable for such territorial nesters.

The plump little males hardly seem able to fly, let alone use the stars and the earth's magnetic field to navigate in the pitch-black darkness over jungles, mountains, oceans, and islands. Soon they will be nesting on the ground: vulnerable to dogs, cats, fox, coyote, skunk, cowbirds, hawk, and especially the blades of tractors mowing this field for hay. It is a miracle that any bobolinks survive, let alone that they return each year from their 12,500-mile round trip journey, one of the most impressive migrations in the entire animal kingdom.

But why do they migrate? A northern ornithologist might ask why do birds fly south but a southern birder might ask the more insightful question: Why do birds leave the tropics where the weather is always warm and the food plentiful? What could possibly prompt nature to select for the costly adaptations that make their migrations possible? Why did birds evolve neurological mechanisms to detect the earth's magnetic field? Why did they evolve the ability to bulk up so radically to fuel their migrations? What are the advantages of such expensive behaviors?

Ornithologists believe that birds' recent migratory behavior evolved as the glaciers retreated and the climate warmed. This would allow them to fly further north. But why would they do so? The answer has to do with time and daylight.

A bobolink that stays near the equator has only twelve hours of daylight to catch food to feed its offspring. But a bird that flies north can gain as much as ten more hours of summertime daylight. Those extra hours are particularly crucial to a ground-nesting bird like the bobolink. A bobolink who nests in New Jersey might be able to raise its chicks from egg to fledging in ten days, but a bobolink who expends the energy to fly to Nova Scotia might be able to do it eight days. This would reduce the amount of time that the chicks and eggs are vulnerable by 20 percent, a crucial margin of safety, enough to make it well worth the effort to fly an extra thousand miles. If it is worth migrating an extra thousand miles to gain a 20 percent better chance of raising your brood, wouldn't it then be worth migrating 6,000 miles to gain a 90 percent better chance of raising your family? Such odds are the stuff of evolution. Nature selects for species as assiduously as a gambler calculating his hand at a blackjack table.

A week later, five more males had arrived. They spent the day sitting in the lindens belting out their mating songs and performing their elaborate courtship flights. Occasionally one would fly down and disappear in the grass, which was already six inches tall. They were starting to establish nesting territories in anticipation of the females' arrival. They defended the nesting territories, but the trees and feeding grounds seemed to be neutral areas. The bobolinks were certainly far less antagonistic than the red-winged blackbirds, grackles, and robins, all edge habitat species that kept all intruders out of their broad, well-defined territories.

But then disaster struck. Six days after the first bobolinks arrived, a truck pulled up below my window and a town official unloaded a large tractor and proceeded to mow a broad path within ten feet of where the males had started to establish their nesting territories. I sent a flurry of e-mails to town hall but to no avail. Evidently they crossed in cyberspace or evaporated in the ether.

But it didn't really matter. The decisions had already been made. This was where the parking lot would be, this was where the walking path would be, and the wetlands would have to be drained to protect a neighbor's driveway. Even though the lot had been purchased to protect nature, even though the literature clearly stated that wetlands should not be drained and paths should be located around the periphery of fields to protect rare and endangered groundnesting birds — priorities were priorities. I tried to make the argument that the town could mow a temporary path that would skirt the periphery of the field during the two-month nesting season but it was to no avail. It would mean that hikers would have to walk an extra half mile, a totally unacceptable inconvenience.

It was clear that this nature reserve, like so many nature reserves, was being designed primarily for the convenience of cars and people not for the sake of rare and endangered species. But perhaps it is too much to ask that humans be temporarily inconvenienced; after all, we have only a few short weekends to recreate — rare and endangered species have all eternity to be extinct.

I walked down the path a few hours after the mowing and it was a strangely empty and depressing scene. Not a single bobolink could be seen or heard. Instead of their joyous calls there was an eerie silence, as if in the wake of a war.

It struck me as ironic. I had spent the previous two months watching the bombing of Baghdad and had noticed that even during the height of the precision bombing you could still hear songbirds sing and watch them flitting through the palm trees and circling above the heads of the embedded journalists. It struck me as odd that in Baghdad we really hadn't been able to get our way despite using awe, shock, and precision bombing, but here in New England we had just succeeded in quieting the birds by building a nature preserve.

The following day the birds were gone. Presumably they had decided that this was no place to build a nest and had flown on to neighboring Plum Island. Unfortunately, I was unable to chronicle what happened in the next crucial four days. I had already made arrangements to observe the bobolinks in Cape May, New Jersey. It was a humbling experience.

Mowing through bobolink territory.

Here I was, a science writer who had always separated my birds into big brown birds, little brown birds, and owls and left it at that. Now I was thrown in with a group of dedicated birders who wouldn't think twice about spending hours waiting for some nondescript little warbler to poke his head out of the bushes. Some of the best field scientists and behaviorists are birders, but for the life of me I can't understand their compulsion to collect life lists.

We only discovered one field out of dozens that had little enough foot traffic to attract bobolinks. I began to realize how fortunate I was to have a pasture right beside my house that attracted more bobolinks per acre than almost any field on the East Coast.

Of course I never would know if those original six bobolinks, who had probably nested in the field last year, had simply retreated into the surrounding countryside or flown on to Plum Island. But perhaps it didn't matter. Six days later a second wave of bobolinks arrived, and there were now eighteen of them. Half called from the graceful arc of nine linden trees that formed a spine down the first field and the other half called from the spine of telephone wires that bisected the second adjacent field.

Six days after that, there were twenty-eight male bobolinks. They had started to establish nesting territories and were spending more time in the grass. But on May 20 the town mowed again and the pattern re-peated itself. The males became very agitated, flying from tree to tree. Only two remained below the trees, and several started to reestablish new nesting territories closer to the periphery of the field. Here they were in greater danger, not just from the more aggressive edge species, not just from hawks that would use the trees on the edge of the fields as launching sites to swoop down on the defenseless chicks, but to an even more pernicious enemy, cowbirds.

Cowbirds have evolved an insidious means of expanding their popu-lations. Instead of laying four or five small eggs and working hard all summer to raise their brood, cowbirds lurk in the bushes on the edges of fields. Then, when an unsuspecting bird like a bobolink is not watching, the cowbird flies out to the bird's nest, pecks holes in any existing eggs, and lays one of its own large white eggs in the middle of the bobolink's

Bobolinks nest on the ground.

nest. The cowbird egg hatches first, and its chick grows faster than any future nestmates. The larger cowbird chick will often brace its back against a smaller songbird chick and shove it out of its own nest. But even if the cowbird does not directly kill its nestmates, the results are the same. The bobolink parents re-spond more favorably to the abnormally large cowbird and run themselves ragged raising the alien chick while their own offspring quietly wither and die.

The cowbirds have perfected such brood parasitism to a high art. They lay over forty eggs a summer, compared to the four or five laid by most songbirds. They parasitize over 200 species that are all declining as the number of cowbirds increases. It is thought that cowbirds evolved this behavior so they could follow bison across the prairies, but they probably evolved the behavior first in South America and then the be-havior allowed them to follow bison herds when cowbirds expanded into North America.

But the bobolinks are not entirely defenseless. They have become a happily polygamous, almost communal species. Several of the females have become attracted to one of the dominant male's territory. The dom-inant male will spend most of his time feeding the chicks of the first fe-

Bobolink food, a praying mantis on a dandelion head.

male to take up residence, but he also provisions the nests of the other, usually less experienced females. All the nests will benefit from having extra females in the territory to keep an eye out for parasitizing cowbirds. But curiously, the dominant male does not seem to impregnate all the females in his territory. They are more than willing to slip off when he is away and steal a copulation or two from a neighboring male. It seems like a successful if somewhat messy form of free love — then again, are there any other kinds of free love?

But something else is going on as well. A younger male bobolink is catching caterpillars for the chicks in one of the dominant male's nests. The younger male is probably the mother's offspring that she hatched the year before. The young male will help feed the chicks and provide another pair of eyes to guard against cowbird invasions. Young females also provide such auntie behavior. Such freewheeling communal behavior is actually the result of hard-nosed kinship selection, a bit of Darwinian insurance to help thwart the insidious behavior of cowbirds. It is another example of the never-ending coevolution of competing behaviors between parasites and their hosts, even though in this instance the parasite and host are in the same family.

By August the males will have molted and look like the drabber females. They will fly back down the East Coast to Brazil and Argentina. On the way they will dine on the rice fields of the Georgia and Carolina lowlands. Prior to the Migratory Bird Act passed in 1925, hundreds of thousands of bobolinks were slaughtered every year in these states, where they picked up the sobriquet "ricebirds." When they island-hopped through Jamaica and Cuba bobolinks picked up the name "butterbirds" for other unfortunate gastronomic reasons. Today they are still killed by rice farmers in Brazil and Argentina, but there are still enough bobolinks to return every spring to our house on the Ipswich Marsh.

Chapter 3
Smith Island

It is June 9, 2003. The sun is finally poking beneath dark, gray rain clouds. It's been raining all week. The sun presents a welcome reprieve.

"Keer, keer." It's the distinct call of a tern. I look up and scan the skies. It's early for terns, strange that they would be out here in the middle of this field, but perhaps they're just flying toward the ocean.

"Pee will willet, pee will willet, cheeriup, cheeriup, thief, thief, thief, haw, haw, haw, mew, mew." I've been had once again! It's just *Mimus polyglottus,* the mockingbird, who, true to his name, mimics birds, cats, dogs, doorbells, even car alarms and the sounds of backing trucks. It's the same mockingbird who fooled me all last summer. I've counted at least ten birdcalls in his repertoire, including robins, blue jay, crows, tern, and willets. He particularly seems to enjoy mimicking the calls of the shorebirds that occasionally visit the wetlands area. It probably makes him sound enticingly exotic to his mates and impossible to compete with to his rivals.

There are fewer bobolinks in the trees and more in the grass than there were a week ago. They have started to pair up and build their nests. A few males have two females in their territories. It is a good thing, too. A pair of cowbirds sit in an adjoining pear tree waiting for a chance to slip in and lay one of their large eggs in an unprotected bobolink nest. The bobolinks will soon have eggs, but it is already late in the season. Yesterday I saw redwinged blackbird chicks sitting on a branch begging for food. This was probably the redwing chicks' last free meal. They could already fly and their parents would soon abandon them to start raising another brood. The "reds" had arrived and started to nest in March in order to have time to raise two families. It is another reason the redwing blackbirds can out compete the "bobs" who don't arrive before May and barely have enough time to raise a single brood before they have to turn around and migrate back to Argentina.

A pile of chewed up apple branches reminds me again of our recent long, cold winter. Mice and rabbits had chewed the branches when they were deep under the snow. It had been a relatively easy winter for prey animals but a hard one for predators. I haven't seen or heard a coyote. While their prey had been hiding below the snow, they had been starving overhead. Many predators had not made it through the winter. There would be far fewer fox and coyotes born this spring, and many more mice and rabbits. They predators will have plenty of food, so the following summer they will trim down the prey animals. Back and forth the two species will go, keeping each other's populations healthy and in check. There is no balance of nature, but a lot of seesawing back and forth in response to biological and environmental changes. It's a good thing, too. It's so much safer to have a self-correcting system than one set in stone. If the preordained stone moves, there goes your system.

At the end of the field I enter a small orchard of fragrant pear and apple trees. Clouds of the most delicate pink and white blossoms float above the canopy. A brilliant flash of orange flitters through the foliage. It is the courtship plumage of a Baltimore oriole, whom the systematicists have recently renamed to just the lowly northern oriole. Too bad, not many birds named after cities now except the St. Louis Cardinals, the Toronto Blue Jays, and perhaps the Mighty Ducks. That's considered to be a real knee-slapper in birding circles.

I am mesmerized by the lush colors and fragrances after the stark monochromatic beauty of this past winter. If there are miracles in this universe, surely the coming of spring is among the most beautiful and inspiring.

But now I must push on. Ahead of me the path rises to overlook Plum Island, Eagle Hill, and the Atlantic Ocean. It is a stirring panorama of drumlins, marshes, creeks, and bays. A lone weeping willow accentuates the verdigris patina of the new spring marsh. It was almost summer in the field behind me, leaves were on the trees, blossoms were in bloom, and the grass was already eighteen inches tall, but spring has just arrived in the marsh. The tender spring shoots of marsh grass are still only six inches high.

A causeway gracefully arcs through the marsh toward Smith Island. The causeway was built in the 1600s to help the Smiths mow the salt-marsh hay. It is an extraordinary piece of engineering. Every month for the past 300 years the full moon high tides have covered the causeway

The causeway, often underwater.

with several feet of water; after the tide recedes, the causeway remains the same. You could drive an Abrams tank over the causeway and it would still stand. Few of our most recently built highways could withstand such extremes.

Several years ago I was in Chatham several months after a new inlet had raised the water level in Pleasant Bay six inches, the equivalent of fifty years of sea-level rise. It was on a calm day with almost no wind. But as the spring tide rose, six-inch waves lapped the shore and quietly started to undermine the new parking lot. In less than half an hour thirty feet of asphalt had collapsed into the Atlantic Ocean. It was a strangely unsettling experience. It would have made sense to see such destruction if the wind had been howling and huge waves had been crashing against the asphalt, but this had been done so easily, almost gently, on a sunny day in spring. It was disturbing to see just how fragile human structures really are against the inexorable forces of nature.

It would never have happened with this causeway. I can imagine the scene as scores of colonial farmers built the structure. They had to re-move tons of boulders from their fields and load them into horse-drawn dump carts, then carefully put them in place. It must have taken several hundred man-hours to just build this short strip.

The farmers used the causeways as staging areas to mow the marshes. Horses could not walk on the soft marsh unless they were shod in specially designed wooden shoes that looked like equine snowshoes. The farmers had to drive the horses, hay wagons, and heavy equipment onto Smith Island, then shoe the horses on solid ground before leading them out the soft marsh. While some farmers scythed the hay, others would drive large circles of locust-tree logs into the marsh. The newly cut marsh grass would be dried on these wooden straddles until it could be harvested as hay. It was a big operation with horses, wagons, and scores of people wielding scythes and mallets. Unfortunately, few people bothered to paint the picturesque scene.

But why all the fuss? Energy! Hay provided food for horses, cows, and sheep. And horses provided transportation, cows provided milk, and sheep provided clothing. Virtually the entire economy of seventeenth-century colonial America was based on hay, and the best and cheapest hay came from marshes.

Salt-marsh hay was superior to upland field hay because it didn't contain weeds and it was virtually free. Farmers didn't have to pay farmhands to clear forests and remove boulders. All they had to do was cut the hay, dry it on straddles, then load it into double-hulled barges and take it to market. The barges were called gundalows because they could be poled gondola style across the marshes during the high course tides and down the creeks to the town landing. From there the hay would be loaded onto hay wagons and transported to Boston. Every major city had a large space near the center of town where the hay was sold and distributed. Boston's area was simply called Haymarket Square. Now it is know primarily as one of Boston's finer subway stops. During the seventeenth and early eighteenth centuries hundreds of thousands of tons of saltmarsh hay were transported from the North Shore and Cape Cod to Boston's Haymarket in order to fuel the growing New England economy. No wonder owning a salt marsh was such a lucrative proposition. It was a little like having an oil field buried beneath your lower forty.

The marshes were so important and so unevenly distributed that most towns designated them as commons areas. In fact, our house used to be called the Old Gate House because town farmers had the right to drive their cattle down Jeffrey's Neck to the broad expanses of salt hay marshes

The Old Gate House.

surrounding Great and Little Necks. The cattle would graze on the commons all summer, then be driven back to their own farms in the autumn. That was where the owner of our house made such a killing. Captain Smith was paid ten cents for every cow that passed through his gate. Presumably he shared it with his fellow townsmen who had milked the cows, shot the wolves, and built the stone walls that kept the cows restricted to the Necks all summer.

I can see what the colonial farmers were after around the edges of Smith Island. Along the upper marsh there is a dark green strip of grass called *Juncus gerardi,* or simply black grass. This was the most valuable grass, used for cattle feed. Beyond it, on land only an inch lower, is a broad plateau of fine green salt-marsh hay called *Spartina patens. Spartina patens* was used primarily for bedding. Fringing the creeks and outer edge of the marsh are the tall spikes of *Spartina alterniflora,* the pioneering marsh grass that made the whole enterprise possible.

It was a far younger process than fossilization that led to our modern fuels like coal and oil. The process only started five or six thousand years ago, about the time that the Egyptians were building their pyramids and Hammurabi was codifying his laws. At the time, this area was all open water still filling in after the retreat of the glaciers. One day a duck flew in from the south and a few *Spartina* seeds dropped off his webbed feet

Juncus *black grass beside* Spartina patens.

and found the mixture of mud and fresh and salt water to their liking. They sprouted and started to send out long underground rhizomes that in turn sprouted in all directions. The *Spartina alterniflora* continued to grow as the sea level rose. Eventually it raised the land just enough to support its cousin *Spartina patens* and the *Juncus* grasses.

Today if you drilled down through the center of this marsh you would retrieve an eighteen-foot core of peat, which would tell you how much the sea level has risen during the past five thousand years. All this was created every year using free energy from the sun and free fertilizer delivered daily by the tides. Now this type of free renewable energy is being displaced by dangerous nonrenewable energy like the electricity being produced at the Seabrook nuclear power plant that I can just see on the far side of the Great Marsh in Seabrook, New Hampshire. Maybe I should stake a small claim on a section of this marsh just in case anything happens to our dependence on the whims of oil suppliers and the obvious dangers of nuclear energy.

My explorations have taken me far out into the center of the marsh. I look up from my photography just in time to catch a glimpse an animal standing several hundred yards away. It must be a fox, but it is not moving like a fox. I crouch down and affix my telephoto lens. If I duck behind the line of trees on Smith Island I will not be visible, but I will probably also lose sight of my prey. Perhaps I can sidle along the edge of the island

If you drilled below this marsh you would get an eighteen-foot core of peat.

keeping some low bushes between me and the animal, which I expect to bolt at any time.

Now I'm within a hundred yards of the animal but I still can't identify it. It carries its rump higher than its head and it seems to have longish ears. Is it a walking rabbit and it just looks farther away than it really is? No, it can't be a rabbit. What would a rabbit be doing out here?

At last I'm within fifty feet of the animal and it still hasn't seen me. Perhaps something is wrong. It's certainly not a fox and too big to be a rabbit. Suddenly it hits me. I'm looking at a newborn fawn still hobbling around on uncertain legs. It is probably only a week old and still too young to run. As if on command the fawn collapses into the marsh grass and disappears. But I have marked the spot and keep my eyes on it while I continue my advance.

As I approach, the fawn becomes even stranger. It has a beautiful reddish brown coat and large soft ears. Yet its ears never twitch and its eyes never blink. Finally I'm sitting beside the fawn and can see her telltale row of white spots. I take several photos, then retreat out of sight to change my film. I return and she is still there. She hasn't moved a muscle; she still holds her head to the side in a characteristic pose. I move in closer to take a closeup and I can almost her say, "Damn, damn. What have I done now? Oh, what did mother say? Gotta keep still, gotta keep still."

Fawn sitting in the middle of the marsh.

It takes all my will not to try to soothe her with words nor pat her elegant head. Now I notice why she probably stood up. She is lying in an inch of water. Her mother probably left her in the middle of the marsh while she went off to graze in the far field. The fawn had sat perfectly still all day until the high course tide flooded her little tussock of grass. I had just happened to look up when she stood up to change positions.

She looks so alone and vulnerable in the center of the marsh where anyone can see her. It seems like the last place a doe would want to leave her unprotected fawn. Wouldn't it be safer to leave her hiding in the forest or the edge of the field? But on reflection I realize that the middle of the marsh is probably not a bad place at all to leave a newborn fawn.

She hasn't moved a muscle.

Deer yard in the Juncus *grass.*

Fox and coyote are not likely to expose themselves by venturing out on the marsh and all our dogs are well behaved. If the high course tide hadn't risen and I hadn't come along, the ruse would have worked.

But I still wondered if something wasn't wrong. It just didn't seem to make sense that a fawn would be sitting in the water alone abandoned by its mother. I scanned the edge of the marsh with my binoculars looking for the distraught doe. But the mother was not there. I walked back to Smith Island and found a deer yard in the *Juncus* grass at the edge of the marsh just beneath a willow tree. I had seen several deer tracks earlier on the causeway.

So perhaps the entire herd would walk out the causeway at dusk to spend the night in the isolated deeryard. Then in the morning the doe would lead her fawn out to this tussock of *Spartina* grass in the center of the marsh. There she could remain all day safe from the prying eyes of fox and coyote, safe from all except the nosy naturalist poking about the marsh to get some pictures.

Having solved the problem, or at least concocted a "just so" story plausible enough to fit the facts, I started to amble back up the path toward home. I took one look back and saw the fawn arise and poke through the marsh as if nothing had ever happened. It was only I who had been transformed by the experience.

Part II

SUMMER

PRECEDING PAGE: *The flats are exposed, and the beach is empty.*

Chapter 4

Crane's Beach

(July 1, 2003)

It is early morning. The beach is still empty. The seven o'clock sun glints off distant sandbars. Last week the winds shifted to the southwest and now the waves of each incoming tide are imperceptibly pushing the sandbars toward shore. It is a subtle process. Each wave stirs up sediments, suspends them in the water column, then redeposits them in ripples a few inches closer to shore. But the results are dramatic. A beach can grow by as much as a hundred feet in a single tidal cycle.

Today there are three sets of sandbars separated by a hundred feet of water. The first has already attached to the shore. The second may arrive in a few weeks and the third a month later. Each time a sandbar arrives the beach will widen by almost thirty or forty feet. Some of the sand will be blown back into the migrating dunes and some will be scoured back into the ocean by storms. But as long as the longshore currents continue to deliver sediment from Plum Island, the beach will endure, growing narrower in winter and wider in summer. The important thing is not to interfere with this mighty ebb and flow of sand.

A long narrow tidal pool lies behind the first sandbar. In a few hours this shallow slough will be the favored spot of children trying to dam its inflowing waters. But now a pair of piping plovers probe in its moist sand and a flock of least tern sit on the dry sandbar looking out to sea.

I am observing the birds from the cab of the Trustees of Reservations pickup truck. The conditions are perfect for censusing these endangered species. The flats are exposed, the beach is empty of humans, and it is warm enough for the plover chicks to skitter below the beach crest foraging for food.

We stop to observe a family of adult plover *Charadrius melodus* probing the moist sandflats for invertebrates. The chicks are a constant blur as they run up and down the beach. Most are only a few days old. These are the ultimate precocial species. The chicks can start foraging a few hours after they hatch, but their metabolism remains slow for a few weeks so they must be brooded. It is an evolutionary trade-off; they can expend their energy on growth rather than keeping warm, but they run the risk of dying from exposure if the weather turns bad. Today the weather is splendid and the chicks look like little wind-up toys, their legs a constant blur below a determined ball of fluff.

It is the piping plovers' love of the beach that has helped make them so endangered. The chicks like to crouch in the tracks of offroad vehicles to avoid heat and detection. The stratagem works. They are all but invisible to anyone sitting behind the wheel of a beach buggy. The results are as obvious as they are fatal.

However, these plovers seem to be doing well. Crane's Beach has been owned by either private owners or the Trustees of Reservations since the early 1900s, so the area never developed a long-standing tradition of beach buggy use. Other beaches experienced pitched battles between environmentalists and beach buggy enthusiasts. One became so heated that the driver of an ORV was seen stomping a plover chick to death in order to protect what he regarded as his human right to drive on the beach.

Piping plover.

Because of its setting and unique history, Crane's Beach has long been one of the most productive beaches for piping plover on the East Coast. Every year it attracts more pairs of adults and produces more chicks than any other beach in New England. The magic number is 1.25 — that's the average number of chicks each pair of plover must raise to hold their population steady. Most years the beach exceeds the number and adds to the entire East Coast population. But it requires hard work. The Trustees have been actively encouraging the plovers since 1986. They do this largely by patrolling and fencing off the nesting area. It's a formula that seems to work. Sunbathers and plovers coexist, with few people noticing the tiny endangered species that scoots between their beach towels to get from the dunes to the shore. The colony consistently produces more chicks than the nearby

Piping plover eggs.

Plum Island Wildlife Refuge that closes its outer beach to sunbathers during the breeding season.

We jump out of the truck and start to census the nests. Erik Johnson, the Trustees assistant ecologist, explains the dilemma. "Up until about a week ago it looked like this year would produce a bumper crop of fledgies. I was out here the day the first adult arrived. It was on March twenty-first, the first of many cold, wet, rainy days of censusing. More adults arrived in the following days and they started to mate and lay their eggs. Each pair laid four eggs. It normally takes twenty-eight days for the eggs to hatch and they hatch every other day so we get to know each bird and when their eggs are due to hatch. This female over here is a very good bird. She stays on her eggs no matter what.

"We have recorded thirty-five pairs of adult plovers that have hatched out thirty-eight chicks. That puts the colony just below the magic number, but there are six more nests to go. That means that there could be as many as twenty-four more chicks, which would make this a record-breaking year.

"But it probably won't happen. Last week disaster struck. We had built these large triangular wire structures called exclosures around each plover nest to protect them from predators. The plover can hop right through the fence but the predators are kept out. But last week, on a

An exclosure.

cold, moonless night a great horned owl discovered the exclosure. He landed on the top and started to hoot and holler and flap his wings. He pecked at the wire until the plover couldn't take it any more. She scooted out of the exclosure and the owl decapitated her with a single snap of its curved, scimitarlike beak. Now the owl has learned that the exclosures mean easy pickings. There isn't much meat on an adult piping plover but like any predator, the owl appreciates fast food. She has probably started to think of the exclosures as a big malfunctioning vending machine — if you kick it in just the right spot, the food will come tumbling out."

But this is serious business for the future of the colony. Four chicks will starve each time an adult is killed. The surviving adult might try to remate and nest again, but the season is advancing. There is hardly time to raise more chicks.

Another predator has learned a similar lesson. A skunk has discovered that if he can avoid touching the bottom two wires of the electric fence that surrounds the colony he can avoid receiving a 5,000-volt shock. Now he is also taking his evening toll of tern and plover eggs. The results are easy to see. Crushed and eaten shells litter the colony and for the past three nights this Einstein of skunks has also avoided the traps set to catch him. Normally you wouldn't see broken eggs in the nesting colony. For years nobody knew where the adults put them but Erik thinks he has the answer: "Last week I saw a tern carrying something down to the water. I ran down to see what it had and fortunately the tern dropped it when it flew away. There it was, an empty eggshell. They must drop the eggs into the ocean so they won't attract predators to their nests. Better than humans. They leave their trash on the beach, which attracts skunks right to the breeding colony."

Now the predators and the terns are in a deadly race to see if this season will be a success or a failure. Every year has its ups and downs. When the Trustees first started their plover protection program in 1986 there were only five adult pairs that produced five fledglings. In 1999 there were forty-four pairs that produced a record eighty-nine fledglings. In 2000 there were forty-five pairs but predators figured out how to thwart the trustees efforts and only twelve chicks survived. Hopefully this year will not be a repeat of 2000.

But there is another endangered species on this beach. Suddenly a cry goes up. A flock of least terns, *Sterna antillarum,* have discovered a school of swiftly swimming striped bass. Now the terns are screaming and jostling for air space over the school. We can just see the shadows of the bass as they swim through the curl of incoming breakers. The bass are decimating a school of silversides. Massacre is everywhere. Twenty-pound bass are driving up from the depths and terns are plunging out of the sky. Fusillades of the silvery minnows are caught in the deadly crossfire.

A male tern plunges into the green waters and emerges with a slender silversides held crosswise in his bill. He flies straight back to the colony and lands with wings stiff and erect in front of a female. Like a tango their courtship continues. He advances, she retreats. He advances, she retreats. She flies out in a wide circle but returns to the enticing offer. If he can provide so well for her, imagine what he could do for her chicks. Of course she doesn't think in such ultimate terms, but the result of her proximate urges is the same. She crouches in the sand and swivels as the male approaches. The male bows his head and extends his neck. Finally the female takes his offer and allows the male to mount and mate. The depression she has made in the sand is the scrape that will serve as their future nest.

It is already late in the season. The tern nests are in all stages of development. Some contain one egg, some two; some chicks are already hiding in tiny copses of goldenrod to avoid the sun. A few have already fledged and will soon fly out to the water's edge to beg for food.

A least tern.

Unlike the more familiar common terns, these least terns like to nest in the foredunes where the beach grass is sparse. Unfortunately, this is also where people like to sit and enjoy the sun. Yesterday I almost stepped on a pair of eggs laid just outside the fence that is supposed to separate the terns from the sunbathers. Today we are back to stretch the fence to enclose several more new scrapes.

The new nests are being built to replace those raided by the skunk. Evidently the terns feel safer laying their eggs in the bare sand near humans than back in the vegetation near skunks. Again we can understand that ultimately they do this to avoid predation, but what are they actually thinking? It would be fascinating to know what psychological and biochemical triggers are shaping these decisions. We know that the female terns do not consciously decide that a certain male will make a good father because of the size of the silversides he presents her. He triggers a series of psychological and biochemical changes that we can perhaps best describe as falling in love. Likewise, when a pair of terns decide to build a new nest they don't consciously decide that an area with little vegetation is less likely to have skunks. That information is locked away in the logic of

Courtship.

their genes and is expressed by psychological and biochemical changes that make the terns feel that such a place is somehow better than another.

People make the same kinds of decisions when they arrange their beach towels for a day in the sun. It just feels right to place them about ten feet from their nearest neighbor. We might have logical explanations for our preferences, but that intuitive feel comes from tiny squirts of neurotransmitters across synapses in our brains.

At the moment my synapses are responding to radically different signals. I am used to common terns, which hover overhead and swoop down to peck at my unprotected skull. In fact, when I photograph common terns I usually wear a hat with a feather in it, so the terns will peck at the top of the feather rather than at the top of my head. One day I made the mistake of forgetting my hat when I left the colony to change my film. I returned to the colony and by the end of the day I had little droplets of blood dribbling down across my forehead.

But least terns are different. They fly straight at you, exactly at eye level, then just before skewering your eyeball, they either skim just over the top of your head or brake and deliver a well-aimed load of excrement

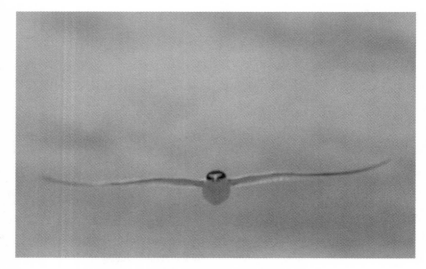

Least terns fly straight at you at eye level, before skimming just over your head.

directly into your face . . . or onto your lens. I find myself crouching for cover and flinching, more like a cowardly war correspondent than a brave natural history photographer.

Ahead of us a more serious problem threatens the colony. A teacher has led a group of students into the dunes and now their feet are blistering from the hot sand. In desperation the teacher has led the children off the path and is leading them single file through the tern colony. The children will probably not step on any eggs but their very presence is a more serious problem. During the day a parent tern has to sit on its eggs in order to protect them from the sun. The children have inadvertently scared the adults off their nests and the eggs are baking just as much as the children's feet. Erik has to intercept the group and lead them back to the path and on to the cooling waters of the Atlantic. It is the most frustrating part of his job as the Trustees assistant ecologist.

I admit I feel some affinity for the group leader. He wanted his charges to have the experience of observing a flock of colonial birds. It is as fascinating as observing chimpanzees or gorillas. Once you understand the meaning of their calls and social behaviors it all makes sense. But the teacher had pushed the envelope. Sometimes good intentions can be as damaging as the foot of an angry SUV driver stomping on the head of a defenseless little chick.

Chapter 5

The Paine House

(July 27, 2003)

It is another long, hot, muggy day. I am walking through our field to the neighboring Paine house. The trip will take me back in time. The house was built eighty years before the American Revolution. But right now, I could not be more in the present. It rained last night and a new hatch of mosquitoes is swarming around my neck, eyes, and face. Blobs of blood smear my upper arms and hordes of mosquitoes are feasting on the few inches of exposed flesh between the tops of my boots and the bottoms of my shorts. It is too hot to wear long pants and I'm wearing high boots to ward off ticks. I will just have to endure these welt-covered knees.

A biologist from Mars might be awed that such a delicate creature as a mosquito could even exist on our planet. After all they are the most fragile of creatures: all feathery wings, legs, and a spindly long proboscis. I have never read a convincing explanation of exactly how such a weightless creature has the leverage to push its proboscis through our thick skins. Yet penetrate it does, along with saliva that keeps our blood flowing nice and liquidy . . . and if I'm unlucky, the virus that causes the deadly West Nile Fever.

I watch a female mosquito, engorged with my rich red blood, fly off my welted arm. She joins hundreds of other mosquitoes buzzing above my head. I look up. Tree swallows are busily darting and banking through the swarm, enjoying a blood feast at my expense. These mosquitoes are helping to transform my flesh into beautiful birds. Did I miss something? Are mosquitoes just God's way of turning human flesh into tree swallows? Or is he just trying to remind me of exactly where I stand in his blood-borne food chain?

Of course, mosquitoes were biting creatures way before humans even invented God. Their ancestors somehow jabbed their sharp little pro-

The Paine House.

boscises through the tough integument of dinosaurs, and their descendents will undoubtedly pierce the skin of whichever species replaces us.

I am wearing my knee-high rubber boots to help protect me against another one of God's little creatures, deer ticks. I have already come down with the high fever and chills of Lyme disease. Tetracycline did the trick, but it was just in time. *Borellia,* the little undulating Gram-negative spirochete that corkscrews through your joints, had not yet released enough toxins to bring on the symptoms of acute rheumatoid arthritis.

I suppose I should be pleased that greenheads, a member of the dreaded Tabanid clan, are not too abundant this year. Perhaps *Bacillus thuringiensis,* the bacteria we spray on marshes to ward off mosquitoes, has also infected the maggots of greenheads. The bacteria will be ingested by the larvae but then turn the tables on the nascent flies and eat the maggots from the inside out. It is not a pretty sight; life is nasty, brutish, and short for these voracious creatures. If a greenhead maggot happens to find his inch-long larval brother burrowing through the soft mud of the marsh, he will not think twice about jabbing his proboscis into his brother's soft transparent flesh. The two will flail about in agony for a few minutes until the first maggot liquefies his brother's innards, then sucks them up through his innocent-looking proboscis.

Perhaps these flies have given me the proper introduction to the Paine House. There will be no idle fantasies about the idyllic life of those who chose to make their living on these salt-marsh meadows. It was hard work. Imagine setting out at dawn among mosquitoes to gather and milk the cows. Imagine carding wool in a smoke-filled front hall with the only light coming through oiled papered windows or a smelly candle made out of the coarse, hard, spluttering fat of a slaughtered pig. Imagine building solid stone causeways over marshes while being plagued by biting swarms of deerflies.

But apparently the deprivations were worth it. When Captain John Smith first sailed through these waters in 1614, he saw the extensive marshlands and noted that the Agawam Indians had already cleared the drumlin uplands for agriculture. Early settlers knew about the potential of salt hay meadows because British farmers had been cultivating them for years. In fact, most of New England's early farms were built near marshes.

Salt-meadow farms were important because they didn't need to be cleared, an expensive and time-consuming prospect. Most New England farmers started out with just a few oxen and a cow so they couldn't afford to clear the land. Land was cheap but labor was expensive, exactly the reverse of the situation they had left behind in jolly olde England. In fact, these salt-meadow farms were so important that by 1690 Essex County had more inhabitants than any other county in Massachusetts except marsh-surrounded Boston.

The farmers' first moneymaking animals were usually pigs. Pigs were cheap and easy to maintain. They could be let loose in the forests to fend for themselves all summer before being slaughtered in the fall to make salt pork. After making a little money from selling pigs, the farmers usually bought more cows. Cows required more work; farmers had to build freestanding barns and cut enough hay to feed the cows all winter. But the small herds provided milk, cream, butter, cheese, and eventually meat. More importantly, cows provided farmers with a way to earn hard cash all year long.

The Paines, Smiths, and Treadwells who farmed these meadows were a hardy lot. Bit by bit, they discovered other things that could be produced to earn currency. Robert Paine built the toll road that led out to the grazing commons on Great and Little Neck. Our house was the Old Gate House, where Captain John Smith collected ten cents for every cow when the cows were driven back into town in the fall. The fees paid for a town herdsman to care for the cattle and shoot the occasional wolf that wandered over Great Neck. Andrew Diamond caught codfish off the Isle of Shoals in New Hampshire, dried them on Diamond Stage in the Ipswich River, then had them shipped to the West Indies, where they were sold as cheap food for slaves.

Other farmers discovered they could cut white oak and fashion it into barrel staves to be used in the West Indies for holding molasses. Some farmers made shoes in their smoky front halls. Later they learned to build ships in these Essex marshes for the lucrative Gloucester fishing fleets.

One enterprising Yankee even came up with the meshugah idea that he could make money cutting blocks of ice out of New England ponds and sell them to plantation owners in the West Indies. Frederick Tudor delivered his first boatload of ice to the Caribbean in 1806. It immediately melted on the docks of Martinique, effectively liquefying all of Tudor's assets. Bostonians started to snicker about "Tudor's Folly." But several bankruptcies later, his folly paid off and New England farmers started spending their winters cutting ice to sell to Tudor, who would ship it to ports from the Caribbean to Calcutta.

Today, you can still see the massive three-foot-thick brick walls of the ice warehouse on Tudor's Wharf in the Charlestown Navy Yard, and you can still drive past innumerable "Silver Lakes" and "Crystal Ponds," once unprepossessing sloughs, renamed in the 1800s to enhance their ice-making image — relics of a New England Ice Age of a decidedly more commercial nature.

But here was the rub. The West Indies had sugar while New England had ice. Yankee ingenuity or not, who wanted to be out cutting ice during cold New England winters while some fat cat plantation owner sipped cool drinks in the Caribbean . . . and his slaves did all the work?

No wonder New England was a backwater. The sugar plantations of the West Indies and the tobacco and cotton plantations of the South were the real moneymakers of the British Empire. In 1640 the New England Colonies only had 40,000 inhabitants while the West Indies had 80,000 citizens, not including their slaves.

The sugar islands were so lucrative that at the end of the French and Indian War the British were going to trade all of Canada for Guadeloupe

An old ice pond.

and Martinique, two islands not much bigger than Nantucket. They were stopped by wealthy British plantation owners, who often lived in London and held powerful seats in parliament. The sugar growers knew that if they let Martinique and Guadeloupe into the empire they would break their own lucrative monopoly on the sale of sugar to British Empire . . . including to the New England colonies.

By 1770, hundreds of New England towns like Ipswich had their own distilleries for converting West Indian molasses to rum to fuel the lucrative sugar, rum, and slave trade. Besides, sugar was becoming more common in New England as tea was coming into vogue. No wonder New Englanders decided to stage a tea party in Boston Harbor. But notice they dumped tea, not sugar, into the harbor. Sugar was too valuable and could be traded for slaves.

So how could codfish and ice really compete with sugar and tobacco? Who wanted to scratch a living out of New England's rocky soil when you could be a plantation owner in the Caribbean? The answer says a lot about economic opportunity and democracy. A few people could make a lot of money in the South, but a lot of people could make a little money in the North. After all, who but a Yankee would think he could make money from selling ice? Perhaps that is what historians mean when the say that New England was made from wits and water, but maybe they should have added in just a soupçon of chutzpah to make the adage ring true. Whatever it was, New Englanders were finally able to accrue enough capital to be finance industrialization.

But it was the basic fairness of the system that made it so successful and long lasting. Even today, nations whose economies are based on a wide range of natural resources are more democratic and distribute their wealth more equitably than countries based on a single natural resource like oil. And it all started on these salt-meadow farms where you could earn a good living if you could put up with hard work . . . and swarms of pesky mosquitoes!

Today the Paine house is owned by the Trustees of Reservations, one of the oldest conservation organizations in Massachusetts. The farm contains three drumlin islands, including Diamond Stage. The house remains as it was when it was built in 1694. For years the Trustees had the Paine House listed as having been built in 1702. But in 2002 the Trustees had a dendrochronologist drill a plug of wood out of one of the oak beams that supports the house. Using advanced microscopic and computer-enhanced dendrochronology techniques he was able to count the tree's growth rings and calculate its age. They discovered that the house was actually built in 1694 and that the oak tree started growing in 1557!

The thing that intrigued me the most about the Paine house, however, was its basement. Tucked into the northwest corner of the cellar was a basement dairy. You could see the brick floor built just above the water table and the benches where Mrs. Paine churned butter. Our house was built around the same time using the same techniques as the Paine house. Did it also contain a basement dairy? After all, Robert Paine had given our land to his daughter, who had married Daniel Smith. Our house had been built by Captain John Smith in 1740. Together, the two farms had been among the most lucrative haying farms in the community.

I raced home and climbed down into our cool fieldstone-lined cellar. Sure enough, there it was, a basement dairy — exactly where it should be, beneath the northeast corner of our house! There was the old doorstep that led down to the dairy. There were the old brick floors and the free-standing wall that never made any sense. There was the geosmin smell of cow manure that I swear I can still smell emanating from beneath the bricks on a hot muggy day in late summer.

I guess I had always imagined that the Smiths lived only somewhat differently than ourselves. Sure, they had no cars, TVs, plumbing, or central heating, but they seemed to have lived reasonably comfortably on income from farm produce and haying.

But it was the basement dairy that really set my imagination on fire. Captain Smith had tended rows of corn, field peas, flax, and barley. The family ate homegrown peas, cornmeal, and salt pork. Hannah Smith had to churn butter in the basement and stand by her friend Mrs. Treadwell on nearby Island Farm, who lost four children in five days to a mysterious throat distemper that swept the colony in 1736. When he died Captain Smith made ample provision to his widow and eleven children, leaving them the house, the wharf, a riding chaise, 200 gallons of rum, twenty-five gallons of brandy and forty-five gallons of wine and three hogsheads of hard cider and four gallons of gin! Yes life really was different then.

After all, these people fought in the Revolutionary War. Robert Paine was the foreman of the jury that convicted several innocent women to hang during the Salem Witch Trial. He later wrote a letter that admitted the jury had made a mistake.

These houses were built only a few generations after the Plymouth Plantation had been settled. Jeffreys Neck, the road we live on, was named for William Jeffereys, a member of the original Plymouth Colony who ran a fishing operation on Little Neck. A cache of rifles was found buried beneath our backyard garden, and arrowheads and finely wrought sinkers were discovered in an Indian mound near a spring on the edge our back lawn. And what about the towering oaks that support these homes? According to dendrochronology, they started growing only fifty years after Columbus discovered America.

So here in my own house I can span the history of our country and go back almost halfway through European civilization. Yet our civilization represents only a millisecond in the life of our species, only the most infinitesimal nanosecond in the history of our universe. To discover just how old this land really is, we must dig much deeper than beneath my basement dairy.

Chapter 6

The Matriarch

A Time of Rapid Change

(August 19, 2003)

It is high tide. Purple bouquets of sea lavender poke just above the surface of the marsh and full heads of marsh grass, *Spartina alterniflora,* tower over me. I am paddling down Fox Creek, so narrow that there is barely room enough to paddle.

Fox creek is the oldest canal in the United States — its construction began in the late 1600s and was completed in 1820 — and arguably the narrowest. It was built for the delivery of logs to Essex, soon to be the largest shipbuilding port in the world. Farmers and shipbuilders had long since stripped these hills of trees, and the Essex yards had turned to lumberjacks in New Hampshire and Maine to supply them with logs large enough to make masts and planking.

But the king's agents had already scoured the North Woods, notching trees reserved for his majesty's navy. It was illegal for New Englanders to use the king's trees to build American-made ships, but many of the king's trees mysteriously drifted down the Merrimac River anyway.

Schooners filled with the king's trees would tie up at the mouth of Fox Creek and either be pulled and nudged down the canal, or their logs would be off-loaded and floated single file down the creek to the Essex marshes, where they would be fashioned into fishing boats; some would even be fashioned into the magnificent clipper ships of the China trade. How many New England trees cruised in and out of Chinese harbors? How many splintered on the reefs of Pacific Islands? One of the former owners of our house owned the largest ship operating in China, and one of my ancestors managed to smash a beautiful clipper ship into a lonely reef in the Pacific. I was once severely berated for my ancestor's transgression when I tried to date the great-great-granddaughter of the family who owned the ship and had a painting of it prominently displayed over their dining-room table. Ancient grudges die hard in New England soil.

Fox Creek.

Today you can still see the canal as a narrow but curiously straight creek that runs under Argilla Road just before you reach the entrance to Crane's Beach.

The remains of an old pier stand above the tidal marsh. During world War II this shipbuilding facility employed 600 men, who worked round the clock in shifts to build landing craft and wooden minesweepers in anticipation of D day. Now the former shipyard has reverted to a lonely marsh.

The land around me looks like the raw material for a golf course. It has hills and ridges, sand dunes and patches of grass. This is not by accident. Golf was invented in Scotland, a land sculpted by glaciers. The sand traps, greens, and rolling hills of the finest golf courses in the world are just a human attempt to recreate the features of a glacially scoured landscape like New England.

But what did this area look immediately after the glaciers had done their work? It was an era of rapid change, when the land was rising up out of the ocean at a rate of nine inches a year and the sea level fell, then rose more than 200 feet in less than 3,000 years. It was also the time

"The king's trees?"

when mammals were getting their first taste of what it would be like to live with humans . . .

It is 12,000 years ago. A herd of mastodons roams slowly over what we now call Ipswich. It is hot and the mastodons are plagued by swarms of biting insects and encumbered by fur evolved for a far colder climate.

The matriarch of the herd has seen the changes. Every summer this thin strand of coast between the ocean and the ice has grown warmer and larger. The Wisconsin glacier, which had expanded as far south as Martha's Vineyard, has now retreated back as far as Maine and is continuing to melt. The mile-thick glacier had pushed this land 150 feet underwater and now it is rebounding from the release of the glacier's weight. The coast is still twenty miles inland from where it will be in 2003. But the matriarch has seen the land rise nine inches in her fifty-year lifetime. Every year she has watched the coast retreat another thirty-six feet. Change is everywhere.

Now the land lies wet and glistening from its overburden of bluish gray mud that was laid down when this land was underwater. The coast is rising so fast that nature has not had time to build up the barrier beaches, marshes, and lagoons that are the features of a rising sea. Without these features biological productivity is low. There are few of the birds, fish, and mollusks dependent on the salt-marsh food chain. The roving bands of Paleo-Indians who have recently migrated into this area

Golf-course geology.

still hunt in the more productive inland areas. As long as the mastodon herd stays on the tundra-covered coast it will be safe from the packs of marauding humans.

The mastodons graze in the shadows of drumlins. These long, elliptical hills formed when the glaciers rode up over immovable outcrops of bedrock. The glaciers had dropped their load of sand and gravel then smoothed the hills as they retreated back north 10,000 years later. Now the drumlins looked like flocks of sheep all facing in the same direction.

Water lies everywhere on this boggy landscape. It cascades off the faces of stranded glaciers; it pools in scattered kettle ponds and collects in glacial lakes hundreds of miles long. It nourishes the tundra grasslands and swells the streams and rivers racing toward the Atlantic. One of the rivers is the ancient Ipswich River that started as meltwater on the surface of a glacier, then plunged down a crevasse to excavate a channel beneath the glacier's solid ice. In its wake the river left thirty-foot-high sinuous ridges of sand that snake across the landscape. Today these eskers are prized as sources of sand for roads, buildings, and concrete.

But now, the mastodon matriarch is faced with a dilemma. She knows it is safer to stay on the coast but the herd is hot and deeply annoyed by the deer flies and mosquitoes. Her own calf is running around her legs, flapping his ears and trumpeting his annoyance.

Water lies everywhere on the boggy landscape.

The matriarch is not too comfortable herself. The insects seem worse than anything she can remember. Finally she decides. She lifts her trunk and smells the air for signs of water. She remembers a glacial lake her mother had led the herd to when she had been their matriarch. The cooling waters had relieved the herd then; perhaps they would do so again.

With a determined air, the matriarch wheels and the herd follows. Excitement ripples through the older animals. They know she will deliver them from their miseries. She has done so before and will do so again. With much jostling and trumpeting the herd trots off into the nascent conifer forest.

A few hours later the herd spots the glacial lake shimmering through the forest of spruce and sedge. The mastodons break into a gallop and splash straight into its cooling waters. Soon they are happily lolling about and spraying each other with their long prehensile trunks. The matriarch's calf cavorts rambunctiously around his mother before spotting a spruce log drifting quietly on the surface. While his mother attends to the herd, the calf swims over to investigate.

Suddenly a great shout erupts from the shore. Six men rise from the log and start splashing furiously toward the infant. Other Paleo-Indians rise out of the sedges and launch more logs from the far shore. The female tries to calm the herd and lead it toward the shore. She knows that on land the mastodons can more than hold their own, but in the water they are slow and vulnerable.

The matriarch tries to regroup the herd but has lost sight of her calf. Two logs have cut him off from the herd and now the men are closing in from both sides. If the infant keeps swimming he may escape his pursuers. But instead he panics and turns back toward his mother.

The first spear glances off his back, but the second strikes and draws blood. Now the calf is trumpeting in pain and mounting terror. The matriarch hears his pleas but can do nothing. Another spear finds its mark and blood spews into the roiling waters. One of the logs draws near enough so the man in front can start clubbing the infant with his stone adz. Each hit draws blood, but it is the one that cracks the calf's skull that finally knocks him into semiconsciousness. A shudder passes through the calf's body and he starts to swim in jerky circles. One of the warriors jumps onto the calf's back and starts to hack great hunks of flesh off his exposed flanks. But the added weight of the man is too much. The infant's body slips underwater and spirals out of sight. On shore the matriarch bellows her rage and sorrow. It is the first time she has encountered humans. She will not bring the herd back to this lake again.

The infant's body drifts slowly toward the bottom, where it decomposes and is covered with sediments. Years later the block of ice that dammed the lake melted and the lake gushed over the flooded plain. Hidden within its load of sediments was the fossilized jaw of the unfortunate young mastodon.

Today I am kayaking near the site of the tragedy. The jaw of the Ipswich mastodon molders in the vaults of the Peabody Essex Museum beside the artifacts of the Paleo-Indians who may have had a hand in its slaughter. The artifacts were excavated from the Bull Brook Site in Ipswich, one of the best studied Paleo-Indian sites in all of North America.

Two thousand years after the tragedy, the land had fully rebounded and the sea level was at its lowest but starting to rise. Blocks of ice still littered the countryside and relic glaciers still clung to the tops of New Hampshire's White Mountains. Water melted off of one of the glaciers nestled into Franconia Notch. It cascaded out of the White Mountains,

Site of the mastodon tragedy? Mating horseshoe crabs.

helped drain the Shawsheen Merrimac glacial lake, and flowed into the Atlantic Ocean as the ancient Merrimac River. By this time the land had stabilized and the river deposited a broad delta of sand laced with inter-braided meltwater streams.

Waves and currents immediately started to reshape the delta into a long barrier beach five miles off the shore of today's Plum Island. Today relic sand from this ancient barrier beach provides material for the modern formations of Plum Island and Crane's Beach, and it provides me with a protected place to paddle my kayak.

So I have ventured back in time far enough to witness the formation of the surface features of this land. It has taken us back to the Ice Ages, and this period of rapid land and sea-level rise. But if I really want to travel back in time, I must dig still deeper into the suspect terrane that still lies below . . .

Chapter 7

In Suspect Terrane

A Billion Years, in One Short Paddle

(August 27, 2003)

It is late afternoon. The sun catches teenagers in midair as they leap off Thunder Bridge near the source of the Ipswich River. I have come up here to get my geological bearings. It will be a daunting task. I am paddling through suspect terrane — that's the term geologists use when they are a little confused. I am more than a little confused about the provenance of the rocks around me. Where did come from? How old is this bedrock? Am I paddling back through time?

It's a beautiful day. Trees arch over sun-dappled waters and long strands of algae sway rhythmically in the swiftly flowing stream. I am drifting along the Bloody Bluff fault line, named not, as you might expect, for the blood red color of its formations, but for the location in Concord, Massachusetts, where it was first described. This geological type locality was also the place where American farmers first bloodied the Redcoats, then the most formidable army in the world. In Massachusetts, we like to pride ourselves on our Revolutionary history, but I feel these rocks hold a far more radical past. I just have to figure it out.

Not far from here, there is a ledge of copper ore being swallowed up by multi-million-dollar resort homes. I discovered it by noticing Copper Mine Road on a local map and following it to its suburban end. In nearby Peabody, Massachusetts, I discovered the Essex Bituminous Quarry. During the day miners blast huge slabs of dark volcanic rock off the vertical walls of the open pit. After hours the pit is haunted by a red-tailed hawk whose long-drawn-out wails reverberate eerily off the steep-sided walls of the artificial canyon. Dark volcanic gabbro overlies ancient formations of iron rich Triassic redbeds. The rocks are crushed into fine gravel then reconstituted into aggregate concrete. Many have been reburied to construct the tunnels of Boston's infamous Big Dig.

The Ipswich River.

But how did these ancient formations get here? It's a billion-year-old story that started off the South Pole, but to explain it we must first take a detour to discuss plate tectonics. By now almost everyone is familiar with plate tectonics. We have seen drawings of the underwater volcanoes of the Mid-Atlantic Ridge. We understand that heat from the earth's interior rises along this ridge and has pushed the American and European plates apart. By merely glancing at the globe we can appreciate that South America and Africa once snuggled together quite comfortably. The same thing is happening to half a dozen other major plates around the world.

But when I was taking "Rocks for Jocks" nothing was quite so obvious. Harvard takes pride in having always had at least one member of the faculty arguing for the wrong side of every major intellectual advancement. At a recent graduation, the dean of students expressed it best: "We want to congratulate you on your recent graduation from Harvard Medical School. You have learned more than any other preceding class. I must caution you, however, that half of what you have just learned will eventually be proved wrong. Unfortunately, we just can't tell you which half."

The Essex Bituminous Quarry.

The geology department handled the dilemma in a more forthright manner. The faculty was hopelessly divided between the "drifters," those who ardently believed in continental drift, as plate tectonics was called in those far more poetic days, and the "transgressors," those who thought it preposterous that something as massive as a continent could glide across the face of the planet. The transgressors believed that all the features of the earth could be explained through a confusing system of flood and droughts labeled oceanic transgressions and regressions. So instead of having one professor teach plate tectonics to one section of the class and another teach oceanic transgressions to another section, the faculty presented the material as a debate and left it up to us students to decide. Remember, this was the antiestablishment sixties. The compromise created an uproar. How could you ace the exam if the professor didn't tell you the right answer? A small minority of us drifters found it edifying to listen to the gears grind as science shifted out of one old outmoded paradigm into a shiny bright new one.

Now almost everyone can agree that during the breakup of the supercontinent of Pangaea, the Atlantic Ocean ripped open and the European and American plates drifted to their present location. However, those

250 million years only account for the last quarter of our planet's history. What happened before? What about those pesky suspect terranes? Where did they come from? To answer those questions we must rewind the geological clock not just 250 million years but a billion years for the bedrock on the sides of the Ipswich River represents three-quarters of the history of our planet!

More than a billion years ago, all the continents of the world started to clump together over what we now call the South Pole. As they clumped together the continents experienced what geologists call mountain building, orogenies. The multiple orogenies raised the Berkshire Mountains, now the backbone of Massachusetts. The continents were forming the supercontinent of Rodinia. But because Rodinia straddled the South Pole, it presented a massive landmass on which glaciers could flourish.

In fact, the world was gripped in an ice age far more severe and long-lasting than the one that gripped New England a mere 30,000 years ago. The Rodinia ice age was worldwide. Glaciers were several miles thick and the oceans almost entirely frozen over. Our planet was little more than a frozen snowball drifting through space.

There were no creatures on land, and the only organisms in the oceans were algae, bacteria, jellyfish-like organisms in the oceans, and stone-like algal stromatolites along the shores of the world. All of these simple organisms were threatened as the oceans froze and photosynthesis diminished. It looked like the fragile little experiment that we call life was about to sputter out and die before it had really begun. But then something changed. About 750 million years ago, rifts appeared in Rodinia and it started to split apart. Volcanoes became abundant, and they spewed globe warming carbon dioxide into the air. Gradually the climate warmed, leading to the Cambrian explosion with its proliferation of a bizarre new assortment of complex life forms.

Rodinia split apart to form the two supercontinents of Laurentia and Gondwanaland. The plates were moving fast and many of them broke into smaller pieces. When this happens, the heavier oceanic plate plunges beneath the smaller broken plate. As it heads back toward the earth's interior, the oceanic plate melts and massive tear-shaped blobs of magma rise back toward the surface. When the magma reaches the surface it erupts as an arc of active volcanoes that delineate the edge of the broken plate. If you want to see an example of this process, all you have to do is fly down to the Caribbean, where the graceful arc of islands marks where the Atlantic plate is subducting beneath the broken Caribbean plate, creating active volcanoes like those that recently erupted on Montserrat and Martinique.

These Caribbean Islands are all being crushed and crumpled together as the plates continue to collide. Eventually all the Caribbean islands will be folded into Central America, thrusting up new mountains in the process of creating a new mini-continent.

The same thing happened with three sets of volcanic island arcs. They separated from the African side of Gondwanaland, then near the South Pole, and were pushed and jostled across the closing Ieapatus Ocean until they crumpled into the eastern side of Laurentia, still below the equator.

Meanwhile, a hundred million years later, the continents had started crushing together again to form the world's most recent supercontinent, Pangaea. By this time, Laurentia with its three little island arcs glommed onto to what would become our own little proto-Massachusetts had moved closer to its present location, almost forty degrees above the equator.

Then, about 250 million years ago, Pangaea started to drift apart as the Atlantic Ocean opened from the south. The Mid-Atlantic ridge did the rest, pushing North America, Africa, and South America to their present locations. Today we know those three little island arcs as the Merrimac, Nashoba, and Avalon suspect terranes. They had formed off of the South Pole, traveled north across most of our planet, and now make up the eastern third of Massachusetts!

So there you have it: a billion years of history compressed into a few miles along the Ipswich River. The next time you find yourself in a discussion about current events you might want to casually insert, "Well of course, most of Massachusetts originally came from the South Pole and migrated to us by way of Africa."

Chapter 8

Hog Island

It is mid-September. Hurricane Isabel is fast approaching North Carolina, but we are still enjoying the golden days of late summer. I have decided to visit Hog Island and Crane's Beach before the storm arrives. I cram my wetsuit and camera into shoulder packs, sling them bandolier fashion across my chest, wriggle into my backpack, and hang binoculars around my neck. I must look like some sort of deranged freedom fighter as I pedal my way down Jeffrey's Neck.

I'm not a gear person by nature. I discovered the wetsuit washed up on the edge of the marsh, my mask and snorkel are old, and my combination road and trail bike is a hybrid no one else wanted at an L. L. Bean clearance sale. But I like to be able to pack for any eventuality. You never know. If the tide is right you may be able to dive for surf clams; if the light is right you may be able to snap a few pics. If nothing else, I'll be able to enjoy an invigorating ride through some of the most beautiful land on the East Coast.

The wonderful thing about riding a bike in September is that you experience the landscape with all your senses. You feel the temperature change when you pass beneath the shade of a hedgerow tree. You smell the wine-sweet bouquet of fox grapes before you even see them. You can pause to sample the flavor of every tart old apple growing beside the road.

It's a shame that someone doesn't harvest all this neglected fruit. I suppose modern palettes are so habituated to the bland taste, large size, and perfect texture of hybridized Red Delicious apples that they will shun the ancient apples that used to sustain our ancestors. But someone should at least mash them into a zesty tart cider.

The marsh is beginning to display its late summer glory. Crimson patches of glasswort stand out against the tawny color of marsh grass

Hog Island in the distance.

heavy with seed. A flock of turkeys have ventured out on a neighboring dock and are now incongruously pecking for food beside the elegant silhouettes of heron and egrets. In the distance looms Hog Island.

Hog Island used to be a name in good repute in New England. It described the highest and best use of many small isolated pieces of land. Scores of Hog Islands were scattered up and down the coast. Now we only have a few; the rest have been replaced by bland generic marketing names.

I remember one former Hog Island on the south shore of Boston. It boasted having the largest fixed rifle-bored guns in the world. The guns could lob a bullet sixty miles north to Newburyport or sixty miles south to Cape Cod. But the neighbors didn't appreciate the giant guns. They were only fired once and blew out all the windows in Hull. Today the former Hog Island houses some of the priciest condominiums in New England. But it has been renamed Spinnaker Island. I suppose Spinnaker is more marketable than Hog.

There were two attempts to rename this Hog Island, once by the town of Essex in 1887 and once again before the U.S. Geological Survey in 1998.

Both attempts were successful but the public never really caught on. Today the Trustees of Reservations still call this Choate Island out of respect to the Choates, but I shall continue to call it Hog Island out of respect to the pigs.

Now the island looms over the marsh in all its spruce covered glory. From its 177-foot peak you can see Mount Agamenticus in Maine, and Seabrook and the Isles of Shoals in New Hampshire. I catch the Trustees launch and head out across a long causeway to the main island. It is like walking back in time, certainly more inspiring than visiting some Disneyesque recreation of "Olde New England." There are no bothersome actors who won't step momentarily out of character, no fulsome gift shops prepared to sell you cranberries grown in Wisconsin or Pilgrim trinkets manufactured in Taiwan. No, this is the real McCoy. You can look through the warped panes of the Choate House and see the same view Thomas Choate enjoyed before the Revolution.

The Choates purchased Hog Island in 1690 for the same reasons the Paines and the Smiths settled on Jeffreys Neck. It was situated on the largest marsh in New England — the largest marsh north of Long Island. The Choates raised pigs, sheep, and cattle and became famous for their mutton, cheese, and butter. They built two miles of causeways. Not to get to town to buy provisions — remember, these were self-sufficient farms that grew virtually all the food they needed. No, the Choates built the causeways for social reasons, to visit friends and transport hay off the island.

The first building you see as you mount Hog Island is the Proctor Barn, a long red barn with double side entrances. It looks more like a barn you might see on a modern farm in Iceland than a traditional old New England barn. But the barn was built in 1778 and was designed with two side entrances to permit twice as much grain to be threshed in half the amount of time.

The second house is the White Cottage. The house was built in World War II during the time when the government was limiting the use of building materials to structures that were already near completion. Cornelius Crane hired two teams of men to work around the clock to finish the building before the deadline expired. Meanwhile, at the other end of Fox Creek Crane's brother-in-law William Robinson had refitted his private boatyard to build minesweepers and landing craft for D-day. Cornelius had met Robinson while sailing in the Pacific and had introduced him to his sister Florence Crane at an elegant homecoming party.

The Proctor barn.

Evidently the marriage had not enjoyed as much smooth sailing as the Pacific cruise. After the war, Florence divorced her wayward husband and had her father demolish the shipyard that so reminded her of William's fickle ways. The White Cottage fared somewhat better. For several years it served as the residence of the island's sheepherder and as Cornelius Crane's weekend retreat.

But it is the Choate House that exemplifies what life was like on a salt-marsh farm prior to the Revolution. The original house was built in 1690 and housed generations of Choates, from Thomas Choate to Rufus Choate, born on the island in 1834. One of their members defended their relative, the lascivious John Proctor, at the Salem witch trials. When the Choates were not busy defending witches they successfully farmed the island for over 200 years.

But in 1917 the wealthy plumbing magnate Richard Crane bought Hog Island. Crane made toilets, so money was of no concern. He had already purchased Castle Neck and built a large Italianate villa complete with extensive gardens and casinos. But he took a different tack on Hog Island.

Crane had purchased Hog Island because it was the first high point of land he saw when he returned from cruising in Maine. He loved Maine, and his family eventually hired laborers to plant a quarter of a million saplings to recreate the feel of a spruce-covered Maine island in the midst of a marsh in Massachusetts.

The Choate House.

Colonial Revivalism was all the rage when Mr. Crane bought the island, so he also hired a local antiquarian, George Francis Dow, to restore the island to its original colonial appearance. Mr. Dow removed several summer cottages but retained the Choate House, which he felt had historic significance. He then restored the Choate House as the centerpiece of his picturesque colonial landscape. It remains today as New England's answer to the so-called English follies, the Greek and Roman ruins the British used to install on their gardens to evoke the past.

But it is still a moving experience to peer through age-warped windows at a view unchanged since colonial days. In fact, you may have already seen this view. Nicholas Hytner chose Hog Island as the setting for his 1996 film version of *The Crucible,* written by Arthur Miller.

Hytner built an entire village and planted a field of corn above the Choate House to recreate the look of Salem Village. He also had to mount a giant live oak tree on a wooden platform so it could be moved to block out a twentieth-century water tower that kept cropping up on the distant Gloucester shore. If you look carefully you can see the same tree jumping from place to place during different shots in the film!

But now it is time to return to Crane's Beach to visit the coastal dunes. The best way to explore the dunes is on sandy trails that snake through

Setting for "The Crucible."

the wild and barren area. I ditch my shoes along the path to better enjoy the feel of the sun-warmed sand.

The wind and ocean have conspired to make this a place of spare beauty. During the winter, storms break through the dunes and blow sand inland across the landscape. Unfettered by the rhizomes of beach grass, the dunes migrate inland several feet a year. I can see where they have almost engulfed a tiny forest of poplar trees. They have also buried farms and several dune shacks. Sometimes you can see where they have buried an ancient orchard. A handful of apples will still cling to a branch that juts just above the sand. The rest of the tree will be buried beneath the sand, thirty feet below.

But once the beach grass can get rooted in the sand it propagates sideways to form a thick network of interlacing rhizomes that can slow the dunes' retreat. The anchored dunes then protect swales covered with velvety carpets of poverty grass. This is *Hudsonia tomentosa,* a heather-like grass that can also be found in deserts and on alpine peaks. It is a succulent grass that lives low to the ground and requires little water and nutrients. In May the *Hudsonia* comes to life and clusters of its brilliant blossoms carpet the swales in yellow. It is a delight to see.

Ahead of me, a winter storm has torn a bayberry bush out of a dune.

Poverty grass.

Its upturned roots are covered with massive growths. They look like they could be a fungal growth, but these are actually nodules the bayberries grow to house nitrogen-fixing bacteria. The nodules protect the bacteria from oxygen, which would normally be lethal to the anaerobic bacteria. In turn, the bacteria draw nitrogen from the atmosphere, where it does nobody any good, and synthesize it into an ammonia-like compound that fertilizes the bayberries.

The bacterium, a *Frankia,* does the same thing in a compost heap or an alligator nest, which is actually just a compost heap in disguise. In the process of fixing nitrogen, the bacteria warm the nest so the eggs will incubate. They also produces acids, which help break down the eggshells, thus facilitating hatching. The warmth generated by fermentation also determines the sex of the newborns. Too much heat produces a male alligator, a good strategy in times of change, and too little produces a female alligator, a good evolutionary strategy in times of stability.

Evidence of how well this bayberry-bacteria symbiosis works is everywhere. Hundreds of warblers and sparrows sit on wire fences built by the Trustees to protect the dunes from foot traffic. A continuous row of partially digested bayberry seeds lies piled up below each wire. The birds have made a point of stopping in these dunes to gorge on the fatty seeds, which will provide them with the fuel they need to continue their migra-

Birds on a wire.

tions. One bird had even been given its name, myrtle warbler, for its habit of feeding on the seeds of bayberry, which used to be known as myrtle. Today we know him as the cedar waxwing, but he still eats bayberries.

The seeds of bayberries can also be used to make aromatic candles. All you have to do is collect the seeds, boil them in water until a thin skin of wax floats on the surface, then dip wicks into the bucket to collect the wax. But be prepared to pick lots of seeds — each candle requires a bucketful.

On the other hand, beach plums require far less work. I find some growing along the path and pluck a bagful to make some of that old New England standby, beach plum preserve. If you make it correctly, the preserve will start out with a lovely sweet taste, then turn quickly and astringently sour, a little like that other standby of this area, a New England winter.

But earth stars are the most fascinating inhabitants of these dunes. I find one rolling on the sand. It looks like nothing more than a shriveled up ball of leaves. But half an hour after it rains this inconspicuous fungus comes alive. A perfect star of basal leaves will fold down to anchor the fungus in place and support a round golfball-sized puffball. Then, each time a raindrop taps the ball, a tiny cloud of spores will spew out to drift out across the dunes and propagate its kind.

Now the sun is setting. I scramble out of the dunes and jog back along the outer beach. I must hurry. The rangers lock the gate at sunset and I want to retrieve the shoes I left beside the trail. I race to the parking lot and plunge back down the path into the dunes. But I have been seen. One of the rangers has seen me duck into the path and uses his bullhorn to announce that the gate will soon be closed. I plunge on. No time to come back out and explain. I reason that it is better to ask for forgiveness than ask for permission. Time is of the essence and at least I will have my shoes.

But now the ranger means business. He hits his siren and as soon as it sounds twenty coyotes howl back in unison. I can see their silhouettes lurking shadow-like in the bayberries. They are far more conducive to good behavior than a truckload of angry rangers. I grab my shoes and hightail it back to my bike, to the amusement of the assembled parking-lot attendants.

As I pedal back home I am accompanied by the full moon and bright red stare of Mars. The last time Mars was this close to earth there were no humans or coyotes to enjoy it. This coast lay beneath the massive hulk of the Wisconsin glacier pushing down from the far north. Who knows who will be here when it returns this close again — or will we be looking back the other way?

Part III

AUTUMN

Chapter 9

Crane's Beach

(September 29, 2003)

There has been a change in plans. I was going to kayak to Plum Island, but when I arrived at Pavilion Beach the tide was rising fast and four-foot waves were crashing into the seawall. Homes and gardens were being sprayed with surf. I decided it would not be wise to paddle in such high waves, alone, in the dark on an outgoing tide — fun, but not too wise.

I spun my jeep around and headed back toward Crane's. It would give me a chance to see the effects of a powerful yet distant hurricane on this barrier beach. I was not disappointed. When I arrived, storm clouds were scudding overhead and shafts of light occasionally pierced through the clouds to illuminate deep green breakers. As they curled over a distant line of sandbars, the breakers shone silver against the wall of deep gray rainsqualls that crouched ominously off Plum Island.

A patch of light splayed bright and curiously over the primary dunes. I jogged down Crane's to investigate. Lobster pots littered the shore and the remains of bayberry plants lay stranded in the wrack line. It was easy to see how the storm had dislodged the lobster pots, but what about the bayberries? They looked so clean and skeletal as they slide back and forth in the outwash. The plants and lobster pots were both the result of Hurricane Juan, which had just slammed into Nova Scotia twelve hours before.

Hurricane Hopefuls were ecstatic; these self-named surfers had been granted their wish. Hurricanes Fabian, Isabel, and Juan had raised high surf all along the Atlantic Coast. Fabian had overwashed Chatham's North Beach, Isabel had inundated the North Carolina, and now Juan was slamming into Crane's.

Waves that had traveled hundreds of miles were now pounding this shore. They happened to arrive exactly during the peak of the highest tide of the season. Now the sun, the moon, the tides, and the waves were

Tearing great hunks out of the dunes.

conspiring to wreak havoc on this shore. The tide was already 11.7 feet high with four-foot waves riding on top. Their combined forces were now tearing great chunks out of the dunes.

As each wave slammed into the primary dune it scoured out a cubic meter of material from beneath the toe of the sandy bulwark. The dune slumped, initiating a cascade of landslides that propagated upward as they loosened the sand above. They seemed to defy gravity. It was hypnotizing to watch. As the sand fell, the landslide inched upward. When the landslide finally reached the top of the dune it started to undermine great clumps of beach grass and bayberry, which slid down the face of the dune and into the roiling Atlantic. The whole process took less than five minutes.

I could see the long-term results of such devastation farther down the beach. Last winter, a series of northeasters had flattened a nearby portion of the primary dune so that now waves could wash unimpeded up over the beach and into the dunes behind. In their wake they left a carpet of new sand only a few grains thick. Clumps of torn-out beach grass stretched out green and straight in the new sand as if they had been dragged through a carding machine.

The results of this process grew around the periphery of the overwash area. Seeds from a past storm had washed in and collected in the wrack line behind the old primary dune. Now a long, green, thin crescent of beach grass marked the furthest incursion of that past storm. Sand had already started to collect around the base of the grass. Storms will blow in more sand and more grass will grow. Eventually this will become a new primary dune, thirty feet inland from the old dune.

There is evidence of this sequence in the newly eroded dune. The face of the dune is striated with black, purple, and white sand. The black and purple beds were laid down when storms delivered heavier grains of garnet and hematite sand. The white silica sand was laid down during times of relative calm. If you drag a magnet across the face of the dune you can watch grains of magnetic hematite jump up and cling to the magnet.

The destruction will provide another benefit. Bank swallows will dig their sixteen-inch burrows deep into the side of the new clean swept face of the dune, and least terns and piping plover will nest in the new vegetation-free washover areas. For it is the amount of erosion that is allowed to happen here that makes this beach such a productive nesting area for such rare and endangered species.

But the storm has also allowed me to witness a rapid deployment in a slow staged retreat. As sea levels continue to rise and storms become more abundant, we will continue to see these events. Waves will batter down the first line of defense, the primary dunes, and the wind will blow sand back inland to build secondary dunes. The system will assume a flatter profile with lower migrating dunes.

Bank swallows.

I can actually feel the process. The wind is blowing a fine mist of sand against my ankles. Some of the sand is also being blown up the gentle western side of the secondary dunes. When it reaches the lee on the far side of the dunes it will drop onto the steeper slope and slide down the leading edge. The dune will advance millimeter by millimeter with each new cascade of sand. The secondary dunes have already climbed halfway up the trunks of a copse of poplar trees, and a carpet of new white sand covers the crinkly dead leaves that make up the floor of this miniature forest.

The secondary dunes will eventually swallow up several more acres of pine and poplar forests. They have already buried some of the farms and shacks that used to nestle in these dunes. Sometimes you can see evidence

Bank swallows nest in newly eroded dunes.

of the destruction. A handful of apples will cling to a branch jutting out of the side of a sandy dune. They will provide the only evidence of an apple tree now buried thirty feet below.

Another way of envisioning this process is to think of the barrier beach as rolling over itself as storms push dunes landward and the beach retreats. Sometimes this process happens so fast that you can see layers of peat in the surf of the outer beach after a storm. The peat is the remains of the marsh that used to grow on the landward side of the barrier beach. The marsh has stayed stationary and the dunes have simply rolled over the top of them.

Cape Cod witnessed a dramatic example of rollover in the nineteenth century. During the summer of 1863, the bleached timbers of an ancient wreck appeared on the outer beach like a ghost rising from the dead. Nobody could remember a ship wrecking on this portion of the outer beach. In fact, the *Sparrowhawk* had not wrecked on the outer beach. It had gone aground inside Pleasant Bay, been duly recorded by Governor Bradford in 1625, then promptly forgotten for 250 years. It had reappeared in 1863, not because it had moved but because Nauset Beach had migrated a thousand feet inland. The barrier beach had rolled over the wreck until it had reappeared on the ocean side.

The same thing has happened less dramatically on Crane's Beach. The remains of a sand schooner, the *Ada K. Damon,* reappeared on the beach

Remains of the Ada K. Damon.

after being buried for almost a hundred years. The western Steep Hill side of the beach has also become rocky as the sand and dunes have been steadily washed away. It is another reminder of how much erosion has already accelerated because of sea-level rise.

If we step back we can see the process speeded up. Twelve thousand years ago the sea level was 150 feet lower and a barrier beach lay exposed five miles east of the Merrimac River. As the seas started to rise, long-shore currents moved this Paleolithic sand southward, where it accumulated in the offshore sandbars off Plum Island. During times of stormy weather, sand builds up in these offshore reserves; during times of calm, currents move this sand back onto Crane's Beach.

In storms, the primary dunes take the brunt of the destruction while dissipating the power of the waves. In the process, the dunes lose sand both to the offshore sandbars and to the secondary dunes. The beach will develop a low profile of narrow beaches and low dunes. During periods of calm the beaches will expand and the dunes will rebuild, but inland from where they were before. The sand grains will become transient members of the sand dunes, sometimes blowing over them, sometimes sliding down their leading edge.

Over time these Paleolithic grains of sand have been plucked from a sunken barrier beach, risen out of the ocean, and been incorporated into the dunes. The dunes have rolled over farms and buildings and filled in

Enjoying the storm.

landward estuaries. During this time the sand, the beach, the dunes, and the vegetation have all acted as a flexible dynamic buffer — a living barrier, able to dissipate storms' power, mitigate the effects of sea-level rise, and heal itself to fight again. But as long as the seas continue to rise, barrier beaches will continue to be just temporary buffers, slowing but not stopping the inevitable erosion of our coasts.

Humans may want to stop this retreat but there is little we can do. Jetties and seawalls only interfere with the dynamic flexibility of this natural system. The best we can do is sit back and enjoy the show. Surfers can hope for hurricanes; birds can enjoy new nesting areas and anglers can enjoy the bounty of estuaries that would stagnate and die if sea levels were ever to reverse or stop rising. The rest of us can just kick back and enjoy the pleasure of watching nature as she reshapes the land and keeps our planet habitable for life.

Chapter 10

Autumn Migrations

(October 15, 2003)

When I awoke this morning the house was shaking with every gust of wind. I lay in bed watching the wind strip leaves off the trees outside our window. One locust had already snapped and others had shed large limbs throughout the back yard.

But by 9:00 A.M. the storm was mostly over. A mass of cold air had pushed the clouds out to sea, leaving cobalt blue skies in its wake. When I went outside I noticed a large, brilliant orange butterfly clinging to a late-blooming dandelion. It struggled to keep its wings pointed into the stiff wind. Suddenly the butterfly let go and sailed willy-nilly back into a tree. There it grasped a leaf before being blown further south. But what appeared to be a random buffeting about was actually part of one of the most incredible migrations in the animal kingdom.

This monarch was part of a phalanx of millions of butterflies using these cold fronts to push and buffet them down the coast. Soon they would arrive in Monarch, Florida, or push on into the mountains of Central Mexico. There, they would spend the winter in suspended animation — the mountains keeping them cool enough so they wouldn't have to eat and the sun keeping them warm enough to survive. The only problem would be global warming or winter storms that can sometimes be so severe that they kill off millions of the fragile creatures before they and their progeny have a chance to migrate back north.

We are just learning how many other insects make similar migrations. This past summer people noticed that an unusually large number of dragonflies were feeding on mosquitoes. At first I assumed that the dragonflies had hatched out of local ponds in response to the rainy weather that had created a superabundance of mosquitoes. But these

Silver-bordered fritillary on hawkweed.

were green darner dragonflies, *Anax junius*. Their carnivorous larvae spend two summers developing underwater before hatching out as the stunning aerialists we are accustomed to seeing. How could they know two years in advance that there would be a superabundance of food? They couldn't. What we had seen were millions of dragonflies that had migrated north in the summer to join those who had already hatched locally. There appeared to be two populations of green darners: one that hatched in the south and migrated north in times of abundance to find new sources of food, and one that overwintered in northern local ponds.

But do the northern layers head south in the fall and does the southern population fly purposefully north? Are the dragonflies being blown in by chance or are they all part of a large migration?

It is easy to get confused. Late one summer in 1998, I was swimming on a beach in Marblehead when I noticed thousands of dragonflies flying out over the ocean. A cold front was attempting to push them out to sea, but apparently the dragonflies realized they were in danger and were fighting their way back to land. I could see the fortunate ones clinging desperately to the wrack line, while others were falling exhausted into the waves. Hundreds would die, but enough would complete the journey to outweigh the considerable costs of their hazardous migration.

There are similar cul-de-sacs up and down the coast. One of them is the southernmost tip of New Jersey. There the dragonflies have to reverse direction and fly back north against the prevailing winds in order to find a place narrow enough to cross the Delaware River. Those that can't beat back against the wind are doomed to perish in the cold waters of the Atlantic.

Of course, if you really want to understand migration you have to turn to birds, and if you want to see birds, one of the best places on the entire East Coast is Plum Island during the migration season. The tip of Plum Island is actually part of Ipswich but it lies on the other side of Plum Island Sound. I could simply launch my kayak and paddle ten minutes across the sound, but the weather has already turned; I jump into my car and drive half an hour up the coast to Newburyport instead. At least the drive will allow me to travel deep into the heart of this marsh, now quickly assuming the tawny colors of autumn.

My first stop is Joppa Flats. It is an extreme low tide and I can look out across a broad expanse of mud flats to the Merrimac River, which exits the coast just south of New Hampshire. This was where the great rafts of logs collected before being guided down the narrow Plum Island River and Fox Creek Channel to their final destination in Essex to be fashioned into fishing boats.

But today I must pay attention to birds. I can hear the distant calls of the greater and lesser yellowlegs. They are easy to tell apart. The greater yellowlegs have a distinctive "tew," "tew," "tew" call, while the lesser yellowlegs get by with a two-toned "tew," "tew" call. Their calls have replaced the gentler "pill, will, willet" of the willets who breed on these marshes all summer.

The two species look similar until they fly. Then the willets diplay distinctive white chevrons that are kept hidden under their flight wings when the willets are feeding. The birds hide their chevrons when they are on the ground to avoid attracting predators. But when they fly the willets' chevrons stand out to help

Willet at rest, with chevrons hidden.

Willet in flight, with chevrons exposed.

them recognize each other and keep their flock together. It is a common pattern among shorebirds.

The yellowlegs are running drunkenly in and out of a larger flock of more sedate shorebirds. The easiest way to tell them apart is by their bills. The dunlins look slightly comical, with their long thick bills that look like they should be on a much larger bird. The dowitcher is larger but has short legs and a long thin bill, while the Hudsonian godwit sports a long, upturned, gracile bill that it probes daintily into the mud flats.

Now a flock of Bonaparte gulls grabs my attention. They are beautiful, flighty, white birds named after a renowned birder who was also a cousin of Napoleon Bonaparte. The sun glints off the Bonapartes' silvery white underwings as they jostle for position over the frothing waters. I see the occasional darker underwings of a less common little gull making his presence felt. Below the gulls, there is a flock of cormorants swimming with their heads underwater. One by one they arch their backs and dive underwater to return to the surface with menhadens held crosswise in their bills. Suddenly the water erupts with the splashes of large fish. They are lunging at the menhaden from the depths of the channel, while the gulls and cormorant dive at them from overhead.

"Must be striped bass," I announce to a group of gathered birders in what I presume to be my most authoritative tones. "If it they were bluefish the cormorants wouldn't stay in the water. The bluefish would bite off their toes." I can't hide my excitement. I have spent too many years as a child and too many lean years as an adult catching bass when they were our only source of protein. I can feel the birders sidling away. Aren't we supposed to be out here watching birds, not salivating over fish? But I do have one ally. An osprey is hovering high overhead, ignoring the birds but plunging unerringly into the maelstrom

Dunlin.

Striped bass with sea lamprey.

to emerge with a striped bass held firmly in his talons. He is probably the only one on the flats today who doesn't want to see a bald eagle soar down the river to steal his fish.

The striped bass have swept into this bay to fatten up for their migration south. Menhaden are a good choice. They feed on plankton and convert it quickly into fat and bone. I ask the assembled group, "How many of you have ever eaten a menhaden? None? Well, how many have eaten a Perdue chicken recently? If you have eaten a Perdue chicken, I guarantee you have eaten menhaden. They are the fourth largest fisheries in the country. They are used primarily for poultry food and also for paint thinner, so the next time you are painting your house you may be painting little bits of menhaden onto the side of your house." The birders turn to concentrate on a godwit.

I decide to continue my investigations at home. If there are striped bass and menhaden on Plum Island, why not on the other side of the sound as well? By the time I reach Ipswich the sun is already setting. I cannot resist walking down to the shore to investigate. Sure enough, the waters around Smith Island are also roiling with fish. I've brought along my fishing rod but it's of little use. All I have is a rusty old plug I found in the marsh. The plug is missing several hooks but it doesn't matter. My reel is jammed so I can't retrieve the lure anyway.

Osprey.

But I have to find out what kind of fish are out there. Perhaps I can just troll the lure behind my kayak. We keep our boats in a neighbor's backyard. It is a good arrangement — everyone keeps their kayaks and canoes in his backyard and it is understood that if you show up with friends you simply grab as many boats as you need and off you go.

As I pull out into the marsh I realize there are thousands of small menhaden swimming just below the surface. I toss out my line and hold my rod under my foot so I can still paddle. "Bang," I have something on. I'm not sure if it's a bass or bluefish until I have it alongside. But now I have a real dilemma. The only way to catch the bass is to land it in my lap. I'm not sure I want a slippery bass flapping around on my lap with sharp extended dorsal spines while I am trying to navigate a tippy kayak. Fortunately, the bass is undersized so I don't have to make any rash decisions. I am able to simply dislodge the hook alongside the kayak and watch the schoolie swim happily away.

It continues that way all night. Fish after fish slam into my lure and I release them back into the marsh. It has become something of a personal rule of thumb. It's only when my rod is broken, my reel is jammed, and I'm in a kayak that I find fish. When I'm well prepared, they are never to be seen.

But perhaps it's just as well. I no longer believe that fish can't feel pain. Scientists have discovered that fish have a nervous system that is just as well developed as our own; their only drawback is that they can't yell out in pain. I always suspected that the old wives' tale was just a convenient bromide to get us off the hook for causing so much pain. But perhaps I'm just getting old and find it easier to identify with the panic-stricken fish struggling on the end of my line.

The striped bass are fattening up for their own migration. They have followed the menhaden into the marsh because it is a few degrees warmer than the ocean waters outside. Each incoming tide is warmed slightly by the mud flats that spent all day soaking up the sun's energy. But the overall trend is down. Every few days the ocean temperature has dropped another degree. The gradually cooling water temperatures make a reliable indicator to trigger fish migrations. But what about the birds flying overhead?

If birds used air temperature to trigger their migrations they could end up in a lot of trouble. A few days ago we had our first frost; next week it will reach seventy degrees. If the birds used air temperature to cue the start of their migration, they could be stranded in an early snowstorm or start their migration too late. But what environmental cue changes as slowly and surely in the terrestrial world as water temperature changes in the ocean? I only have to look at the setting sun to get my answer.

The sun is setting half a minute later today than a few days before. Day length is the environmental cue on land that increases gradually and reliably in the spring and decreases gradually and reliably in the fall. The decreasing daylight in autumn is the cue that triggers the bird's hypothalamus to send a signal from the brain to the pituitary gland, which will in turn release hormones into the blood to make the bird start to seek out fatty foods and become more and more restless until they migrate. I can see that restlessness in a flock of shorebirds along the edge of the marsh. As sunset approaches, first one, then another, then the whole flock flies out over the marsh, returns, then sets out steadily south. In fact, most of the shorebirds left the Arctic Circle when the light first started to diminish in the summer. What I am seeing is the tail end of the tropical migration and the beginning of the arrival of birds that will spend their winters in New England.

Finally it is fully dark. A pack of coyotes howl on the far edge of the marsh and are answered by more coyotes on a nearby hill. I lie back in my kayak and stare up at the stars. Mars still shines reddish orange in the southeast and Venus is rising out of the eastern ocean; I feel like I am suspended in a hollow sphere of stars. The infinite depths of the autumn sky are reflected star by star on the mirror-like surface of the waters below my kayak. The mighty arc of the Milky Way bridges the horizons above and below. The flashes of a high-flying jet are lost in the austerity of the night. It is only when the craft is directly overhead that I can hear its diminutive roar. It breaks the atmosphere, a crude reminder of humans' pervasive domination of our planet.

But now the roar ceases, to be replaced by the piping of invisible birds. Migrating across the starry vault, they call to each other to maintain their flock. Out here it does not seem so incredible that they can navigate by the stars. They — as well as I — are suspended in a vast sphere of revolving constellations. They — as well as I — watch the stars roll out of the eastern ocean and settle in the west. We watch the reflections of the stars as they wheel majestically overhead. Only the North Star remains motionless at the apex of the sphere. The birds only have to set their course a few degrees to the side of the unerring star to accurately align their flight.

Eventually these birds will see the Southern Cross rise slowly out of the south. They will know they are close to their homes in the jungles of Amazonia or the pampas of Argentina. It gives me a great feeling of independence to know that I can jump into my kayak and paddle out under my own power to catch my dinner. Imagine what it must feel like to be able to fly from one hemisphere to another under your own power. Imagine what it must be like to feel at home in North and South America. Imagine flying more miles in a single year than most people travel in a lifetime. And all of this is accomplished by a little ball of fluff you can hold in the palm of your hand, a little ball of fluff with the brains to migrate 20,000 miles and the physiology to make the trip and reproduce at its end. It is surely one of the most majestic miracles of the universe, a miracle we have come to take largely for granted.

Chapter 11

Man and Coyotes

(October 30, 2003)

Our first frost occurred on October 3, the first snowstorm on October 22. Today it is in the seventies so this must be an Indian summer, a period of warmth that arrives after the first frost of the year. Clusters of ladybugs sit in the sun preparing to crawl into our house to start their group hibernations. The chirring of crickets is back up to last summer's tempo after declining to a few quiet chirps per second. Even some unwelcome mosquitoes have hatched out in this unseasonably warm weather.

I am trying to finish some writing when our dog starts to bark. This is not just his usual "I'm bored and want to rejoin the pack," bark. He is clearly barking at something that has him annoyed — if not frightened. My wife yells, "Come look. There's a coyote in the backyard circling Keiko!" Perfect excuse to quit writing.

I rush to the window. Sure enough, a large healthy coyote is dodging and feinting at Keiko. The coyote has a bushy large tail with a thick black tip that looks as if someone dipped him tail first into an inkwell. The coyote lunges in, then feints to the side. Keiko barks furiously and charges to the end of his chain. The coyote knows that Keiko is chained and seems to enjoy his frustration. The tails of both canids are outstretched and wagging, but the coyote looms far larger than our harrier, a slightly oversized breed of beagle.

I decide to not take any chances. I run out and drag Keiko inside. The coyote retreats a few steps but ignores my yells. It is only later that I have a chance to ponder the encounter. The coyote did not seem to be stalking or hunting Keiko. If he had, I fear my intervention would have been too late. If mating were on their minds, the coyote would have been sorely disappointed. Keiko has been neutered. The coyote never came closer than ten feet to Keiko, but if they had actually come in contact would their play have turned rapidly into violence? Would instincts have taken over and would the coyote have snapped onto Keiko's domesticated little rear flank?

The incident puzzled me all night. Finally it dawned. The coyote had not been trying to attack Keiko; neither had he been intent on playing or mating. He had simply been trying to draw Keiko away from his bone so he could sweep in from the side and steal it. Since it was three in the morning I decided not to reveal my discoveries to my recumbent wife. But at dawn I rushed to the window. Sure enough, there was the same coyote sitting beside our doghouse, happily chewing on Keiko's half-eaten bone.

I grabbed my camera and raced outside. The coyote seemed unperturbed by my bathrobe and nightshirt. Rather than run off, he walked boldly in my direction before turning to trot down a side path leading into the phragmites. I followed him to a temporary den only twenty feet from our driveway.

Coyotes, *Canis latrans,* or barking dogs, have been spreading from the northwest for more than a century. Now there are coyotes in every state in the union except Hawaii, and judging from their occasional sightings on the offshore islands of Nantucket and Martha's Vineyard it is just a matter of time before they figure how to purchase an airline ticket or swim from Alaska. The first coyotes trotted into New England fifty years ago. But evidently a lot had happened during their expansion. Eastern coyotes now have wolf genes and are a lot more like small wolves than western coyotes. When they were first discovered in New England they were called "new wolves" or "coydogs" until the mammalogists got their act together and decided to call them eastern coyotes.

But what had brought this creature into our yard in the middle of the day? I had seen him many times before: once sitting in the field, once trotting casually beside our car, this times threatening our dog. What made him so bold? The rest of the coyotes in the neighborhood made their presence known only by howling at night, not by hanging around in the day. Was he rabid? He certainly looked healthy and well fed. Also, the other coyotes in the area seemed to stay in packs; this coyote was a loner. Was he an adolescent coyote newly on his own? Did he feel emboldened because the town hadn't mowed the field for the first time in 300 years so he could now hide out in the tall grass close to people's backyards?

Coyote glowering at my intrusion.

What I was actually seeing was an artifact of mankind. Coyotes have been steadily expanding eastward precisely because they have been persecuted so long in the West. For the past hundred years humans have tried to exterminate coyotes. Today government agents and hunters slaughter over 400,000 coyotes a year.

In Maine, hunters use snares to slowly asphyxiate the animals. The snares often cut off the coyotes' jugular veins but their carotid arteries keep pumping, so the coyotes' brains fill with blood until their vascular system bursts from the high pressure. The hunters call these animals "jelly-heads" because their brains are mushy from so much semicoagulated blood. It is a little like suffering from migraine headaches, while being choked for three days, before dying from a massive stroke. But there is another problem. Snares cost a buck apiece, so it is easier to leave them set in the forest rather than retrieve them at the end of the season. The result? The snares act like ghost traps, continuing to strangle deer, moose, bear, elk, and Canada lynx, a species deemed threatened in the United States. The bodies of the dead animals probably attract more predators that also die in the ever-constricting loops.

So what have been the results of all this slaughter? Coyotes have continued to expand and increase their populations. In fact, killing coyotes stimulates the remaining coyotes' reproduction and increases their numbers. The reasons are many. Coyotes have spread across the nation largely because we have extirpated the wolves, bears, and mountain lions that used to dominate most of the country's prime habitats. However, these animals were the pinnacle predators of their particular niches. They never had any predators above them in the food chain, so they never evolved an innate fear of other animals, including humans. Coyotes, on the other hand, always knew that if they didn't play their cards right they would end up as somebody's dinner. Over time they evolved to become the resourceful generalists we know today. They have learned to exploit humans but not trust them, a trait they share with cockroaches, pigeons, and rats.

Today coyotes use their brains, their vocal cords and their highly evolved social behavior to bring down game much larger than themselves. They also use these higher faculties to out-compete and avoid being eaten by higher, more specialized pinnacle predators. Sound familiar? These are the same strategies early humans used to become the resourceful animals we are today. Like us, coyotes are vocal, crafty and wary. They use a simple form of language to help them hunt and maintain relations both within and between packs. These generalist strategies allow them to thrive in habitats from deserts to grasslands to forested mountains. Recently they have started taking kindly to the suburbs and have even been seen ducking in and out of traffic in the Bronx, probably looking for a pick-up game with the Yankees, an equally unsavory predatory species.

Coyotes have also learned to eat almost any kind of food, whether it be fruits, nuts, and berries or mice, rabbit, dogs, cats, pet food, deer, and elk. They can live singly, in pairs, or in packs, like humans, and like humans, their social versatility has also allowed them to expand into almost every niche in North America.

So if you kill off coyotes, why doesn't that reduce their populations? The reasons are food and predators. The prime predators of young coyotes are old coyotes. Coyote mothers don't even let their mates enter the dens for fear they will kill the pups. This is a common practice among predators. A male lion will often kill a female's pups to induce her to ovulate so he can be sure he is perpetuating his own genes. Perhaps the female even knows she stole a copulation when her mate was away. So what is her all-around most conservative mating and parental strategy? Not to trust any male. She simply uses growls and body language to tell her mate to in essence to "Just regurgitate on the doorstep and go away!"

So when humans slaughter adult coyotes they do two things to increase the number of pups. First, they remove their prime predators, and second, they insure that there is more food available. Because coyotes tend to increase in an area until they are close to its carrying capacity, killing adult coyotes only increases the amount of food available to the pups in their parents' hunting territories. Under normal conditions, on average only one or two pups survive into their second year, but when the adults are being poisoned or hunted, on average four to six pups make it through their first summer.

Perhaps the well-known theologian Reinhold Neibuhr had coyotes in mind when he was sitting in the small Berkshire town of Heath trying to come up with a new prayer. The Berkshires are known for their coyote problems, and Heathens are known for drinking. But I still like to think he was watching coyotes when he wrote the prayer, part of which has become a standard at Alcoholics Anonymous meetings:

> God grant me the serenity
> To accept the things I cannot change;
> Courage to change the things I can;
> And wisdom to know the difference.

Anyway, I have determined to ignore the revulsion I feel when I see coyotes chase down and dispatch a deer. I will accept that they are here to stay. I will put the cats inside, pick up Keiko's bones, and will continue to enjoy the frisson of fear that travels down my spine when I hear their wavering calls at night and watch them cavorting in the dew-drenched meadows at dawn.

Chapter 12

Ipswich Clams

(November 19, 2003)

During the summer, I discovered a favorite swimming spot on Crane's Beach. It's a narrow stretch of deep water that whooshes between the beach and an offshore sandbar island. The island is a curiously stable feature. It seems to mark the point at the end of a standing wave, where competing currents build up the sandbar but maintain the narrow gorge that separates the island from the beach. The gorge speeds up the currents so that twice as much water passes over the bottom. The result is a jet stream of water so replete with oceanic plankton that it fuels its own superenriched little ecosystem.

But it is only by diving at low tide that you can appreciate this unique world so different from the din of people enjoying the temporary nature of the island above. Even at the lowest neap tide the gorge is still over my head and fast flowing. Sand dollars litter the bottom. These flat little echinoderms seem immobile but they are in fact crawling on tiny little feet. I once helped set up an underwater camera that automatically took a picture of sand dollars every few minutes. When speeded up the echinoderms looked like comical little Volkswagen beetles weaving back and forth through early morning traffic.

But something else was also attracted to these fast-flowing waters. Occasionally I could see sprigs of algae sprouting incongruously from the bottom sediments. By treading water on the surface and stretching my legs toward the bottom, I could just reach the algae. After rooting around with my toes for a while, I discovered that each sprig of algae was attached to a huge clam lying concealed just below the sediments. These were surf clams, *Spisula solidissima,* who lie buried with only their siphons exposed. Only in this fast-moving gorge were the currents powerful enough to deliver the food the clams needed to fuel their rapid growth.

After loosening the sand with my toes I dove down and tried to dislodge one of the six-inch clams before I ran out of air or the buoyancy of my wetsuit pulled me back to the surface. After several tries I finally had one of the monsters gaping on the surface.

The clams were only four or five years old, but they were already too large to carry comfortably in my hands. The only thing to do was to stuff them into the top of my wetsuit. In no time, I had attracted the attentions of a curious striped bass who circled around my dangling feet, no doubt wondering where all the delicious smells were coming from. After a few more dives I emerged from the waves and strode confidently back up the beach. I assumed the clams made me look like some well-muscled surfer with a chestful of rippling pecs. But the image was somewhat marred by the scatological sounds of the clams opening and closing and the large quantities of water cascading out of the bottoms of my wetsuit. Moving about so close to my heart, the clams felt more like large friendly cows than inert little mollusks destined for dinner.

I had to set such philosophical considerations aside, however. The sun was setting and I had to bike back home to shuck the clams. That was the hard part; I would have to shuck the clams when they were still alive. Unlike quahogs or oysters, which shut down tight, surf clams emit audible gasps and gurgles when you try to slip a knife between their shells to snip their fleshy abductor muscles. It was quite unnerving but had to be done. The clams have an organ called a crystalline stylus that looks like an inch-long piece of flexible plastic tube. I wanted to observe a stylus while the clams were still fresh and alive. Finally I found a greenish gray stylus jutting out of one of the clam's stomachs. I removed the stylus and placed it under my microscope. It was teeming with spirochetes, bacterial relatives of the same spirochetes that almost gave me Lyme disease after I was bitten by a tick last June. But these were good spirochetes. Under magnification, I could see them spiraling through the stylus, helping the clam to digest planktonic cellulose; if I had contracted Lyme disease their bad cousins would have been spiraling through my joints, happily digesting the tissues of my elbows, knees, and back.

In research circles, surf clams are famous for their large transparent eggs. Much of what we know about how human eggs and embryos develop comes from research on the eggs of this uncommonly large mollusk. But surf clams also serve a more mundane purpose. Most of the clams we eat in fried clams and quahog chowder are actually the fleshy mantles of surf clams that line their shells. These "strip clams" lack the succulent stomachs of the real thing, *Mya arenaria,* the soft-shelled clam. But that's OK, surf clam mantles taste almost as good. After dissecting the clams, I would remove their feet and mantles and simmer them in milk along with onions, pepper, and flour. Afterward I would shape them into little balls, roll them in bread crumbs, and bake them until they were succulent golden brown clam cakes that would only last for a single meal.

But now, by late November shellfishing has assumed a more serious mien. I am standing on the town dock waiting for about a dozen commercial shellfishermen to return from digging on the Ipswich flats. Trucks from the Ipswich and Essex Seafood Companies are usually here to meet the boats, but today it is too close to Thanksgiving. This is traditionally the time of year when the demand for shellfish is low but the supply of clams is high because the weather is still good for digging.

Today is the same. One by one the small aluminum boats arrive at the docks, but no trucks: "Charlie says Ipswich is buyin', but you have to deliver the clams yourself."

"Hell, I might just as well take these clams and bury 'em back out in the flats."

"Life's a bitch, ain't it?"

The shellfish warden arrives to measure the clams. Not much to worry about. Ipswich clams are much larger than most East Coast clams because they grow so fast in just the right mix of fresh and salt water, mud and plankton.

The fishermen separate their catch into cutters and steamers. The cutters will be shucked and used for fried clams. The steamers are smaller and will be kept in their shells and used as steamed clams, prepared essentially the same way the Agawam Indians used to cook steamers in their traditional rock pit clambakes held along this same river thousands of years before the Europeans arrived.

Fried clams are a more recent invention. In 1916 Chubby Woodman, an Essex potato chip baron, decided to roll some leftover clams in bread crumbs and toss them into a vat of boiling potato-chip oil. The result was magnificent. The breadcrumbs provided a crunchy salty-sweet outer coating that perfectly complemented the tender stomachs of the clams

The clam flats.

within. Woodman's Seafood Restaurant was off and running and remains a North Shore landmark today.

But it also turned out that Ipswich clams have just the right *"gout de
terroir,"* the taste of the soil, that wine connoisseurs go on and on about
when describing their pinot noirs. Here in Ipswich, we speak more of
"le gout de la mud," for it is the fine silty clay of the many rivers that
debouche into the salty waters of Ipswich Bay that gives the clams their
distinctively sweet taste. Cape Cod clams, in contrast, grow largely in
sand and are therefore whiter and cleaner and have more of "le gout de
la mer," sought out by the gourmets of the steamed clam world.

Four years ago the shellfish warden opened the Ipswich River to
clamming for the first time since the eighties. Before that, the river had
been closed for over fifty years. The unfortunate truth is that today the Ipswich clam industry relies as much on tradition as on reality. The total catch
of Ipswich clams is way down from when the small town supplied more
clams than any other community in the country. Now most clams distributed from Ipswich come from the muddy flats of the Damariscotta

The Ipswich Shellfish Company.

River in Maine rather than from Ipswich. But their reputation lives on. I did a quick check on the Internet and found Ipswich Clam clubs in places as distant as California and the Carolinas. The clubbers would meet regularly to gorge on clams shipped in from Ipswich by FedEx. I suppose they would spend most of their meeting debating whether Ipswich or Cape Cod clams are better. As a recent émigré from Cape Cod, I remain steadfastly neutral on the entire subject.

Chapter 13

The Deer Hunt

(December 5, 2003)

It is December 5. I am walking behind the dunes of Crane's Beach. The path leads me through a variegated tapestry of sere and muted colors. The sandy swales of poverty grass give way to golden carpets of pine needles offset by patches of pale green lichen. Earth stars sit seemingly dead and discarded in banks of moving sand. Birds flit through a thick understory of shrubs and bushes. Deer tracks crisscross back and forth through the poverty grass — but none of the tracks are new.

The Trustees of Reservations held its annual controlled deer hunt a few days before. A handpicked group of hunters had slowly marched down the length of the four-mile-long island, culling the herd of female deer.

Most of the shooters were old hands at this hunt. Back in 1985, they had been selected from an initial pool of 250 hunters from the ten towns surrounding Ipswich and Essex. A hundred of the original shooters dropped out when they learned about the rigors of qualifying. The rest had to attend a yearly three-hour presentation that explained the science, the rationale, and the rules of the hunt.

They also had to meet at a local rifle range to test their shooting skills. From this pool, the Trustees had selected the twenty best marksmen, whom they called the night before the hunt, to tell them to be at the beach at 5:30 sharp. If they arrived too late they would miss the hunt. They could only shoot two antlerless deer. This was difficult for some hunters who had been brought up to believe it was unchivalrous to kill a young deer or doe. Besides, it was only the old antlered bucks who made good trophies. One hunter couldn't help himself and bagged a large antlered buck on the first day of the season. The trophy still hangs in his den, but the hunter has never been invited back.

Why all the care and secrecy? The reasons go back to 1985 . . .

Pine forest behind the dunes.

Fred Winthrop, the new director of the Trustees of Reservations, called his secretary: "Hi Trudy, any messages?"

"Only one, sir."

"Anything important?"

"Well, it was from a Mr. Cleveland Amory."

"You mean the guy who wrote *The Proper Bostonians?*"

"Yes, sir."

"Well, what did he have to say?" Winthrop could hear the hesitation in Trudy's voice. He had just started working for the Trustees and she didn't know him very well.

"Well, sir, he says you're a son of a bitch!"

Winthrop laughed; almost from the beginning of his tenure at the Trustees he had been on the receiving end of hate mail and death threats. It had all been over that damn hunt out on Crane's Beach. His predecessor had tried to open the dunes for hunting back in 1983. It had been an unmitigated disaster. Animal activists had swarmed into town from as far away as New York and Connecticut. They had threatened to infiltrate the hunt and throw themselves in front of the deer. The television stations had had a field day. All the Boston stations had on-the-spot coverage as the hunt approached. Eventually the Trustees had to cancel the hunt just a day before it had been scheduled to begin. The grandfather of the animal rights movement, Cleveland Amory, made several trips to Boston — some to protest, some to raise funds. On one occasion when he was asked what he would do if he were given absolute power to rule the world, he answered, "All animals would not only not be shot, they would be protected — not only from people but from each other. Prey would be separated from predator . . ."

Separating prey from predator had actually been the cause of the problem, but it had happened more than a hundred years before. When Thoreau was writing about New England prior to the Civil War, the largest wild animal he reported seeing was a muskrat. He noted with astonishment that a farmer in a neighboring town had actually seen a living deer! Where were all the deer, turkey, beaver, moose, and bear that had been present in colonial times? They had disappeared as settlers had cleared the land for farming. Sixty percent of the land had finally been cleared for agriculture. This was the peak of New England's pastoral era; her fields harbored grassland birds like meadowlarks, bobwhite, and bobolinks, normally found more on prairie lands. However, the Civil War had opened up the Midwest and exposed New England soldiers to the riches of these new areas. New England farmers abandoned their boulder-studded fields and moved to the more easily tillable lands of the Midwest. In their wake, New England forests started to regain their losses. Today New England is forty percent open and sixty percent forested, exactly the reverse of what it had been in 1865. It had been a massive experiment. In an era concerned about deforestation, it is instructive to remember that New England is still the largest area on our planet that went from intense agriculture back to reforestation. The process took most of a century. During the last stages, deer, turkey, beaver, moose, and bear started to return, some on their own, some by restocking. Today they continue to return at an accelerating rate.

But during the intervening years, other things had changed as well. Wolves and mountain lions had been extirpated to protect livestock. Now there were no pinnacle predators to keep deer in check. The prey animals had also changed. Deer, fox, and raccoons had started to adopt behaviors that allowed them to thrive near humanity. They became nighttime and crepuscular creatures that learned to raid trash cans and family gardens. They became something between a wild animal and a tame one — not feral exactly, but something more like pigeons than the wild animals they once had been.

Deer were now far more abundant in suburban areas than in rural ones. Nowhere was this more evident than on Crane's Beach in the seventies and early eighties. By 1983 there were 340 deer in an area that could only support 60 deer. The results were obvious. There were no wild flowers, bushes, or young sapling trees on the 1,400 acres behind Crane's Beach. Deer browsed adult trees as high as five feet off the ground. You could see under the branches of the trees all the way to the ocean. The deer themselves were stunted and emaciated. The herd had consumed all the readily nutritious food and had started to graze on beach grass, which was causing severe erosion on the outer beach. Adult deer were ripping Christmas wreathes off of people's front doors; fawns couldn't reach enough food to make it through their first winters. The Trustees of Reservations, normally a quiet organization that went about its business buying and protecting threatened land, was spending all of its time trying to deal with its deer problem.

When Fred Winthrop was hired in 1995, he decided the Trustees had to meet the problem head on. For the first time in its history, he had the Trustees hire a professional biologist employed specifically to evaluate the deer problem and come up with a viable solution. Rob Eblinger took stomach samples, collected field data, made aerial surveys, used sharpshooters, and scoured the country to find other areas that had similar problems. Slowly and carefully he started to build a case for a controlled hunt. But it was difficult to build an argument on wildlife management and habitat arguments alone. They always seemed to pale beside the visceral response of seeing hunters slaughtering large, charismatic animals that looked like Bambi — even though Bambi might be half starved from malnutrition.

But the real problem turned out to be that Bambi was spreading a crippling disease sometimes even fatal to humans. Lyme disease had first been identified among people living in three towns surrounding Lyme, Connecticut. It had been described as a new oligoarticular disease that causes arthritis-like symptoms. Later it was discovered that the same disease had been described a century before among people living seasonally on Naushon, a privately owned island off the southeast coast of Massachusetts. Researchers were even able to locate specimens of white-footed mice that had been collected on Naushon Island in the 1800s and left to languish in Harvard's Museum of Comparative Zoology. The mice still had DNA from the spirochetes for Lyme disease in their desiccated blood systems.

The implications of the Naushon mice were fascinating. Lyme disease was not a new disease at all. Instead, it had probably always been present in the New England woods but as a disease of ticks, not of people. It had only been in rare instances like on Naushon that the disease had been able to jump from ticks to humans. Because the disease had been rare, doctors had misdiagnosed its high fever and arthritis-like symptoms as the result of the flu or simply aging.

That had all changed with the explosion of deer populations in suburban areas. Suddenly ticks were able to get up off the ground, multiply, and be spread by these large herbivores. Deer ticks, who had been aptly named *Ixodes dammini,* before the DNA boys changed their names to *Ixodes scapularis,* were the reservoir of the disease, but deer were the vectors that spread the disease more effectively to humans. When deer ticks hatch from eggs in the leaf litter, they climb six inches high on blades of grass, where they can only be picked up by humans on their shoes or ankles. But when the larval ticks drop off mice, they climb several feet high on stems and branches, where they can grasp onto deer or humans. It is easier to pick up a tick by brushing up against it on a branch than by stepping on it in the grass. The little arachnid is also far more likely to find the warm safety of your hair or armpit if he starts off on your upper body rather than on the bottom of your shoe. But the main problem is that deer provide a large source of blood, which induces the ticks to mate and spread the spirochetes that cause Lyme disease. More deer mean more ticks; more ticks mean more spirochetes and more Lyme disease. It is this fascinating coevolution of two behaviors, climbing six inches as a larval tick and two feet as a nymph, that allows the spirochetes to jump from mice to humans and be spread by deer, which don't develop the disease themselves.

This chain of events started to become apparent in Ipswich when the first two cases of Lyme disease were diagnosed in a couple living near Crane's Beach in 1980, and when hunters started finding new kinds of ticks on their freshly killed deer.

The locus of this outbreak was Argilla Road, the main thoroughfare leading onto Crane's Beach. It was the same road where people had been

The view from Argilla Road.

complaining about deer ruining their orchards and ripping the Christmas decorations off their front doors. But Argilla also had another thing going for it. The road had been settled around the turn of the century by a group of doctors from the Massachusetts General Hospital. The descendents of these original doctors were still a savvy lot. In no time at all they had medical students from Harvard, Tufts, and the MGH working on the problem. The researchers quickly discovered that sixty-six percent of the residents living adjacent to the reservation had Lyme disease and that thirty-five percent of all the people living on Argilla Road tested positive. Their symptoms included headache, fatigue, fever, meningitis, paralysis of the seventh cranial nerve, memory loss, arthritis, and heart stoppage. One child had even been previously diagnosed with schizophrenia until doctors discovered her symptoms were also the sequela of Lyme disease.

News of the findings caused instant conversions. One grandmother who had spent every weekend on the village green protesting against the hunt became an overnight convert when her son refused to bring his children to see her as long as deer were still in the neighborhood.

The debate raged on, eventually pitting two old-line New England organizations against each other. One was the Trustees of Reservations, the other the Massachusetts Society for the Prevention of Cruelty to Animals. Both organizations had been established around the turn of the century and both were run by boards of well-connected, old-line Yankees. The owner of the *Boston Globe,* William O. Taylor, happened to sit on both boards. Gradually, he had become convinced that Rob Eblinger and Fred Winthrop had done their homework and that a controlled hunt was the only effective way to reduce Lyme disease and the deer that spread it.

Mr. Taylor arranged for a private meeting to see if the boards of the two organizations couldn't work out their differences before the deadline. Rob Eblinger presented his findings, outlined why the hunt was necessary, and described how it would be conducted. But in the end, the two groups couldn't find common ground. The following day, Bill Taylor resigned from the MSPCA and wrote an editorial in the *Boston Globe* endorsing the hunt.

The straightforward, utilitarian, species-centric argument had won. In the fall of 1985 the hunt went ahead with little fanfare and a lot of secrecy. Both trained professional marksmen and regular hunters participated at different times. After two years the sharpshooters were dropped and the hunt continued, reducing the herd by a half every year. By 1990 the herd was at its carrying capacity, the average number of ticks per deer had declined fourfold to less than seven ticks per deer, and the number of ticks on mice had declined fivefold to an average of less than one tick per mouse. Most significantly, the number of new cases of Lyme disease had started to decline — but it would persist as an endemic human disease.

The experiment had proven that the rapid overpopulation of deer had resulted in the outbreak of Lyme disease and that the hunt had started to bring the disease under control. The results had not been lost on animal rights advocates, who stopped mounting major campaigns against controlled deer hunts.

The fascinating aspect of this entire episode is that it started as an environmental problem in the developed world rather than as a human disease in a developing country. Usually when we read about such emergent diseases they are animal diseases that jump from animals to humans, like Ebola and AIDS, which jumped from apes to humans; SARS, which jumped from civet cats; or flu, which jumped from ducks, chickens, and swine to humans. In all these cases, the disease makes the jump because of human practices, whether eating bush meat in Africa or raising pigs and ducks too close to humans in Asia.

These diseases gain widespread attention because they are so fast-moving, contagious, and often lethal in our age of global travel. But mostly the diseases attract the world's attention after they have started to kill humans. Later, scientists are able to work backward to trace the origins of the diseases back to their animal reservoirs. On Crane's Beach the process worked in reverse. Ecologists noticed the effects of deer on the environment before they realized the effects of deer on humans. Had the situation been reversed, the hunt would have probably moved ahead with far less public opposition. But the process did have its advantages.

The hunt on Crane's Beach was based on a similar hunt on Great Island in Wellfleet, where an entire herd of deer had been eliminated. As would be expected, Lyme disease disappeared as well. Some would say the experiment had been carried out to its logical conclusion.

On Crane's Beach, however, scientists learned that if deer are reduced to the carrying capacity of the land, Lyme disease will slowly retreat back into its animal reservoir and become primarily a disease of white-footed mice once again. Not a bad compromise for a society that increasingly wants absolute guarantees of safety for an animal like ourselves, which seems to be genetically attuned to put the rights of its own species ahead of the rights of other animals.

Part IV

WINTER

Chapter 14

Why the New England Patriots Beat the Miami Dolphins

A Tale of Snow, Cold, and Mitochondria

(December 8, 2003)

Winter slammed into us early this year. It arrived in the guise of a howling two-day blizzard that left us buried under three feet of snow with five-foot drifts. The only animals to truly benefit from the storm were the small herd of deer on Crane's Beach. The Trustees canceled the hunt after only eight deer had been killed.

The winds were blasting sand grains three feet off Crane's Beach and blowing them horizontally toward the dunes. They were strong enough to abrade anything in their path. Surf walkers from the old Life Saving Station used to hold wooden shingles in front of their faces to protect their eyes and skin when they patrolled the beach, looking for ships grounded in the pounding surf. By the end of the night their ears and noses would still be full of the sharp crystalloid grains of sand. Last year I found a wine bottle that a careless picnicker had left on the beach. The sand has blasted it a beautiful frosted green, as efficiently as the marble on a newly cleaned cathedral.

The remaining deer hunkered down in copses of scraggy pine, waiting for the weather to pass. With the hunt canceled they will be under little stress until the coyotes start to drive them through the deep snow. The coyotes have displaced the foxes that used to live on the island and they seem to present a more robust habitus with each generation. Are we witnessing one of those periods of rapid evolution when coyotes are being selected for larger and larger size? In time, will both their size and behavior match those of the wolves that used to fill this predatory niche?

A troop of harbor seals cares little about the howling winds. They simply lie low on edge of the river flats protected by several layers of

Ipswich winter.

blubber. Only their moist eyes cake up with blowing sand but that will wash away as soon as the seals plunge back into the roiling waters. As soon as the weather breaks the seals will be at it again, riding ice floes playfully down the river. Seals are also mammals that adapted rapidly as they evolved from living on land to living in the nutrient-rich waters of estuaries. Like most animals that have gone through such rapid evolution, they still have to return to the terrestrial dangers of their ancestral environment in order to mate and bear their young.

Mice and voles must be rejoicing with the snow overhead. Now they can build their labyrinths of tunnels well stocked with caches of dried seeds and hay while they wait out the winter beneath thick layers of insulating snow. While their predators must contend with the howling winds and freezing temperatures above, they will be enjoying the warm humid pleasures of their subnivean homes.

Autumn has winnowed down the numbers of summer birds; now mostly gulls and crows remain. They wait out the storm, knowing that in the morning they will be able to satisfy their gluttonous appetites on the rafts of clams, crabs, and perhaps even some lobsters washed ashore by the violent storm.

I try to walk down one of the familiar paths but make little progress through the mid-thigh snow. I use a ski pole to scrape the coating of wind-blown ice off an official sign. It reads, "Nudity prohibited."

Harbor seal.

"Damn." I decide to return home.

By afternoon, the snow has stopped falling and the New England Patriots are playing the Miami Dolphins. I turn on the tube to watch the fans reveling in our prototypical New England conditions. The owners have allowed the fans to enter the stadium early so they can clear their seats. Instead they craft high thrones, or simply sit in sleeping bags nestled into three feet of snow. As the game winds down into another gutsy Patriots victory, the stadium erupts into a spontaneous show of light and snow.

Delirious fans throw fistfuls of powdery snow into the air and the lights glitter off the silvery puffs as they erupt in time with the music pulsating through the stadium. Across the nation, TV viewers must be marveling at the bizarre exuberance of this new species of humanity evolving like seals and coyotes in New England.

A few days later the journal *Science* runs an article that might help explain New Englanders' strange affinity for snow and cold. The story goes back almost a billion years to when a species of blue-green bacteria parasitized an early prokaryote. Over time the bacteria became symbiotic, then essential to its hosts. Today the ancestors of those early bacteria make up mitochondria, molecular batteries that break down glucose to produce the energy that powers muscles and basic metabolism.

Things proceeded for hundreds of millions of years of evolution. Those initial prokaryotes evolved into eukayotes, then vertebrates, and

The winter marsh.

eventually into early humans. The humans first appeared from ape-like ancestors about a million years ago. All the while, mitochondria was being passed through the female line independently of DNA, so we know that every human today is descended from a single female dubbed the mitochondrial Eve. Everything was fine until about 65,000 years ago, when two lineages of humans decided to radiate north out of Africa. Eventually they ended up in northern Europe, Siberia, and traveled across to America near the close of the last Ice Age. But during those years nature selected for humans whose mitochondria were less efficient than their relatives who had remained in the tropics. Instead of efficiently producing chemical energy, the mitochondria produced more heat. Although the change made the mitochondria less chemically efficient, it made the new lineages of humans more successful at living in northern climates because it increased their production of body heat. But one result of this change is that when the two lineages eat a high-calorie diet today, the lineage that stayed in Africa is more likely to produce oxidative damage and fatty deposits that underlie such diseases as heart attacks, stroke, Alzheimer's disease, and Parkinson's disease.

Of course, people from northern regions have their own host of diseases. One of these is seasonally affective disorder, which is more common in northern climates and almost nonexistent on the equator.

The implications of this theory are that everyone is adapted to living in a particular climatic zone and that moving to a different zone will cause stress — like the Miami Dolphins flying north to lose twelve–zip to the mitochondrially challenged New England Patriots. And we thought it was all because of Tom Brady and our brilliant defense!

Chapter 15

Coping with Winter

(January 19, 2004)

It is frigid. My daughter and I are cross-country skiing across the frozen surface of Ipswich Bay. The ice is over a foot thick but we are not taking any chances. It is low tide so the ice rests safely on the mud, unseen below us. Giant upturned blocks of ice rise eight feet above us on the edge of the marsh. The tide has dropped some of the one-ton blocks of ice onto erratic boulders, and now they lie in massive jumbles above the frozen bay. On all sides the ice spreads out around us as far as the eye can see: toward Rowley, toward New Hampshire, and toward the Atlantic Ocean. Snow blows across the ice and covers our skies, like in a scene from a PBS film about traversing the frozen Arctic. We would not be surprised to see a polar bear stalking us from the far side of a pressure ridge.

The entire East Coast is locked in a frigid mass of cold air that envelopes the world from the North Pole to New England. Nantucket is running low on fresh food because the cold has iced in its ferry. In New Hampshire, presidential candidates should probably be talking about global warming, but who would listen when the wind chill is twenty below freezing, day after day after day?

Unfortunately, these days of unrelenting cold may not mean that our fears about global warming are unfounded. Quite the reverse, they may mean that global warming has increased to the point where the Gulf Stream may be in the process of breaking down, which could cause the North Atlantic Ocean to freeze as it did during the last ice ages.

Let me explain. Global warming does not necessarily mean that every place on earth will get warmer. On the contrary, ice samples show that during past periods of global warming the northern hemisphere actually became colder. Normally the Gulf Stream carries warm salty water north,

Salt pan.

which moderates the East Coast and helps to deflect the jet stream. This also tends to break up the high-pressure masses of cold air that flow out of the Arctic. After the Gulf Stream carries salty waters north, the waters cool, sink, and return to the tropics as part of a deep-water counter-current. It is the sinking of this cold salty water that helps draw the Gulf Stream northward.

But now, with global warming, glaciers are melting faster, releasing more fresh water to the surface of the North Atlantic. These less saline waters no longer have the density to plunge downward, pulling the Gulf Stream north. Climatologists' great fear is that the Gulf Stream could stop flowing altogether and that northern waters would then freeze and drive us back into another Ice Age. In the interim we can expect to continue to experience the hotter summers and colder winters we have experienced in recent years. Last winter we had twelve successive blizzards and record-breaking amounts of snow; this month we are well on our way to the second coldest January since meteorologists started keeping records in 1888. If this is the pattern we can expect in the future, what is likely to happen to the plants and animals that have adapted to a more moderate climate?

We can see some of the potential problems as we ski back home. Last winter a thick blanket of snow lay over the landscape. We can still see where mice and voles gnawed the lower bark of trees while they trundled back and forth, warm safe and hidden in their subnivean tunnels. This year there has been little snow and predators have had the advantage. But what about less visible creatures?

My daughter stops to investigate a thicket of goldenrod that juts above the snow. Each plant has a bulbous growth halfway up its stalk. Last summer a female goldenrod fly, *Eurosta solidaginus,* inserted a single egg into the soft freshy stem of the fast growing plant. The egg hatched and the larva excreted chemicals that caused the golden rod to form this hard, tumor-like gall. But just before it became trapped, the larva chewed an escape route to the outermost wall of the gall, then retreated back to the gall's center to prepare for winter. It had to excavate this escape tunnel as a larva because when it emerges as an adult in spring it will have no chewing mouth parts.

But this is not all the larval fly is programmed to do. It also produces an alcohol, glycerol, and a sugar, sorbitol, that both lower the freezing point of its blood. All these adaptations have been selected to operate within specific parameters. If the temperature stays too high or falls too low, the larval fly will die. If the changes happen slowly, enough evolution may be able to adapt, but if they happen too rapidly because of something like the Gulf Stream stopping, the species could go quickly extinct.

Another example lies ahead. The remains of a nest of wasps, *Dolicho-vespula maculata,* clings to one of the European linden trees planted in the field. I walked by the tree all summer and often heard the wasps buzzing but never realized how close I was to their basketball-sized nest. It is only now with the leaves gone that we can see it up close. Last spring, a single female wasp came out of hibernation, scraped some fibers off a dead branch, and mixed them in her mouth to make a slurry of wasp spit and wood fibers. This she proceeded to extrude from her mouth into long strips that quickly dried into a grayish white paper. Soon she had constructed a walnut-sized nest into which she started to lay her eggs. When the eggs hatched, her daughters continued to construct this multi-layered structure of recycled paper. As the extended family grew, it remodeled its nest by chewing down the inner walls, remasticating them with spit, and plastering this onto outer walls. We could even see lighter and darker strips of paper where the workers used different kinds of wood. Air trapped between the paper walls added insulation. By shivering their flight muscles, the wasps could keep the temperature inside the nest at 80 degrees while the temperature outside was a cold as 40 degrees. But as the outside temperatures dropped still lower, most of the workers died off and only the queen crawled under the bark of a tree to hibernate.

Honeybees, *Apis mellifera,* employ even more sophisticated and human-like strategies to make it through the winter. Instead of hibernating or migrating, they are the only insect to store enough high-energy food to maintain their hive and body temperature all winter. It is an amazing feat, made possible because of each individual's total subservience to the welfare of the hive. This is only possible because each worker is sterile and identically related to each other. It is in all of their best interests to make sure that the fertile queens survive. There are no selfish genes in a honeybee hive. It makes for an amazing lifestyle.

During the course of the year the female workers carefully select and feed certain offspring special royal honey to one of their chosen sisters who will then mate and become the next fertile queen. In the late spring, the female workers evict the old queen, their mother, who flies off with with a retinue of 20,000 to 30,000 loyal workers to seek out a new place to establish a hive. During the next four or five months, the workers store more than 200 pounds of honey and pollen to support the hive through the winter.

During the winter the bees keep the hive at ninety-five degrees Fahrenheit. They do this by clustering together, but the bees on the outside of the cluster cool faster than those on the inside, so the outside bees try to crawl back inside. When the external temperature lowers further, the outside bees can only force their front ends into the cluster, thus plugging up any holes and sealing in the heat. But eventually the external temperature drops so low that the outside bees have to shiver to stay warm and generate heat for the hive. Meanwhile the bees inside the cluster become too hot and force their way to the outside of the cluster. So there is no central command to control this coordinated response. Pheremones are not exchanged; the queen is not in charge. Each individual simply re-

Harbor seal.

sponds individually to the temperature, but their responses keep the cluster within a degree or two of ninety-five. It is much like Adam Smith's invisible hand: Each little capitalist simply goes about his selfish business and the result benefits the entire hive.

As winter ebbs, however, selflessness becomes the norm. Then individual workers streak out of the hive in search of the first trees and flowers to bloom. Time is of the essence. The bees have to start making honey as soon as possible, so most of these early forays are suicide missions. The snow in front of the hive is strewn with the inert bodies of bees who streaked too soon, lost body control, and plunged into the snow and froze. Hundreds of bees may die in these suicide missions but eventually the day comes when a few bees fly out to a nearby poplar or red maple, and one gathers up some pollen and returns alive to the hive. The workers gather around to smell and palpate our hero bee. She then performs a highly stylized waggle dance that inform her hivemates of the distance, location, and value of the blooms.

So what will happen to honey bees if the climate changes? Killer bees provide an intriguing clue. When northern honey bees prepare to make their winter flights they spend time shivering to raise their body tem-

Late winter storm.

perature enough so they have a chance to fly out and back without losing body control and crashing. Killer bees simply fly out of their hives and drop dead in the snow, victims of a climate too cold for their genetic programs.

And what of the polar bears we imagined we could see prowling across the marsh? A thousand miles north, on Hudson's Bay, infant polar bears are dying because their mothers can't hunt enough seals to regain the 400 pounds they lose to hibernation each winter; without that needed fat, the mothers stop nursing and abandon their cubs. The reason the mothers can't find enough seals is that the pack ice now breaks up two weeks earlier than it did twenty years ago, so the newborn ringed seals that the polar bears count on for food have moved out to sea. The pack ice now melts two weeks earlier every time the average Arctic temperature climbs another degree — and it is expected to climb three to five degrees in the next fifty years. And what will happen to the hapless polar bears? They will probably go extinct as the wooly mammoth did before them. We don't even know exactly what direction global warming will ultimately take us, yet we are already seeing the results and they do not bode well for the future of our planet, other species, or ourselves.

Chapter 16

The Magnolias of Ravenswood Park

(February 18, 2004)

It is February 18, and I am half-skating, half-sliding on Old Salem Road, once the main thoroughfare between Gloucester and Salem, now a broad icy trail wending its way through Ravenswood Park. Snow flakes drift out of the sky and wood chips fly as a pileated woodpecker augurs a six-inch feeding hole in a dying hemlock.

On a warm rainy July day in 1806, the honorable Theophilius Parsons was driving his carriage along this same road when he spotted some creamy white blossoms flowering on a tree covered with shiny green oblong leaves. Surely this flamboyant bloom was not of New England stock! The famed jurist was an amateur botanist and had traveled widely. Could these possibly be the same magnolias he had seen in the Carolinas?

The following week Theophilius returned to the swamp and carefully cut samples of branches, leaves, and blossoms in all stages of development. He marveled at the perfect cup-shaped form of the lovely three-inch flowers and happily inhaled their delicate, lemony-sweet fragrance. Even their leaves exuded a tangy, anise-like aroma.

Later that week Parsons wrote to his great friend the Reverend Manassus Cutler:

I think you have traversed the same woods herborizing. Did you discover it? If not, how long has it been there? . . . It is so near the road as to be visible even to the careless eye of a traveler. Supposing the knowledge of this flower growing so far north might gratify you, I have made this hasty communication.

Your Humble Servant,
Theoph. Parsons

Parson's discovery created quite the stir. Soon curious naturalists started making annual pilgrimages to the botanical shrine. Henry David Thoreau

Sweet bay magnolia.

walked from Salem to Gloucester to marvel at the seven-foot trees, and the head of Harvard's botany department, Asa Gray, rode his horse up the coast to see the northernmost example of *Magnolia virginiana*, the sweet bay magnolia tree that normally thrives only in the bogs and swamps of the deep South. How did the trees get there? Were they a relic from a former population? Did they arrive as seeds in the crop of a migrating bird? Why had they escaped notice in an area that had been settled since 1635? One anonymous author speculated darkly that the trees were not actually wild at all, but had escaped from a cultivated specimen!

The town fathers were so impressed with the discovery that they changed the name of the Kettle Cove section of Gloucester to Magnolia. Another author offered the snide observation that the selectmen made the name change at exactly the same time the town was trying to attract the tourist trade of the mid-1800s.

To understand why Parson's discovery made such a stir, we have to look at the state of the sciences at the time. Botany and horticulture were in their prime in the 1800s. Cities like Boston, New York, and Philadelphia were building grand arboretums. Researchers like "China" Wilson from Harvard's Arnold Arboretum were the biotechnologists of their time. They traveled to the furthest reaches of the globe to gather exotic seeds, which they would grow in specialized botanical gardens, then introduce to newly wealthy American businessmen who wanted to re-create in America the country homes and lavish landscaping of European estates.

I had a great-great-granduncle, known famously for founding the Arnold Arboretum and infamously for screwing up the taxonomy of the crab apples, who spent several days and nights clambering to the top of Mount Hokkaido in Japan in order to collect seeds from two species of magnolia. On his return, he planted the seeds at the Arnold Arboretum and in his garden in Brookline. Both species bloom in May and June. Evidently the trees hybridized on their own, for today the fragrant blossoms, magnificent foot-long leaves, and silky smooth bark of the sil-ver parasol magnolia hybrid grace parks and gardens throughout North America and Europe.

But back in Gloucester things weren't going so well. In 1871 an illus-trator for the Arnold Arboretum reported that when he visited the swamp there were plenty of the fifteen-foot trees. But he also noted, somewhat ominously, that it was easy to find the specimens because of the trails left by boys who stripped the trees of their leaves and branches and sold the fragrant blossoms at the Magnolia train station. By 1914, only two tiny plants remained; the rest had been stripped for sale and then succumbed to the ravages of winter.

In recent years, however, the story has taken a more positive turn. In 1980, a group of horticulturists from the Arnold Arboretum surveyed the magnolias and found several of them in bloom. They collected 935 of the brilliant scarlet-red seeds, sprouted them at the arboretum, then trans-planted them back into the swamp as young seedlings. By 1995, many of the seedlings had survived and the population was in good health.

But in 2002, ecologists from the Trustees of Reservations, who had been given the park to maintain, found the trees to be significantly in de-cline. Deer had been using the trees as rubbing posts to help remove the new spring velvet from their antlers. Many of the trees had snapped off at the scarred areas where the antlers had abraded them. When new shoots of magnolia sprouted from the old roots, the deer would browse those down as well. Some of the old root balls went back at least 200 years and were probably also losing their vigor. Another problem was

Lady slipper or, more properly, squaw moccasin.

the fast-growing red maple and yellow birch, which were forming canopies over the magnolia, blocking the light they needed for photosynthesis and growth.

Today ecologists from the Trustees of Reservations have returned to solve the problem. They are in the process of building an eight-foot-high fence that will surround a five-acre portion of the swamp that contains the remains of the once flourishing magnolia stand. They will also cut down some of the taller trees that are shading the magnolias. It is an exciting project. The magnolia will return. The swamp will once again delight generations of curious naturalists and botanizers. It will remind us of a time 200 years ago when New England was thrilled to make the discovery of this botanical delight in the midst of our seemingly inhospitable climate.

As I was writing up these notes later that night, I chanced to look out the window. Mars was still glowing red above the horizon as it had all year. But this time I knew two tiny robotic explorers were crawling across its surface taking samples, drilling rocks, snapping pictures. Like Theophilius

Parsons in Gloucester and "China" Wilson in Asia, they are looking for evidence of life in a seemingly inhospitable climate. They have already discovered that water flowed across this planet, which had once enjoyed a warm and springlike climate and harbored many volcanoes. A satellite flying overhead will soon start using radar to look as much as five kilometers below the Martian surface to search for water. Most cosmologists now believe that that if you find underground aquifers associated with volcanoes, you are almost certain to find life. When we do find life on Mars or on another celestial body, will it be similar to the bacteria that thrive around our own undersea volcanoes, or plants the size of magnolia and animals the size of horseshoe crabs? If so, will we send other robotic explorers to bring seeds back to earth so we can sprout them at the Arnold Arboretum and return their saplings back to a terreformed Mars? Will future tourists make pilgrimages to Mars the way Henry David Thoreau traveled to Gloucester to view her sweet bay magnolias?

Chapter 17

A New View of Nature,
A New View for Ipswich

The New England Biolabs Campus

(February 23, 2004)

It is a beautiful late winter afternoon. The cold has finally broken; it now
edges above freezing on a daily basis, and I can see the sun setting almost
a minute later every day. When I was here last winter it was a far different scene. It was Washington's Birthday and we were in the midst of a
raging blizzard. It was obvious it was going to be the last major storm of
the season, so I had decided to go out and take some pictures in the teeth
of the full storm. Most photographers wisely wait for a storm to pass before venturing outside to take pictures of newly fallen snow against
cerulean skies. What do I do? I go out in the teeth of the storm when
there is hardly any light, snow clogs up your lens, and you risk frostbite
every time you reset your shutter speed.

My plan had been to cross-country ski to Appleton Farms to see if
couldn't get some shots of snow-covered cattle hunkering down against
the storm. But the wind was too strong and the snow was too deep. I decided to return to the nice warm safety of my car. It would be just too
damn embarrassing to be found dead and rotting in a snowdrift next
spring.

I headed out onto the highway. I was alone. Only a solitary snowplow
had passed this way before. I decided to swing into the driveway of New
England Biolabs. Halfway up the mile-long drive I spotted a red-tailed
hawk sitting on a fence post only ten feet from my car. I eased the car to
a halt and quietly rolled down my window. With my telephoto lens I was
able to squeeze off a full-frame shot of the snow swirling about the head
of the handsome bird. He didn't move. I shot several more pictures, then
finished the roll without his moving. He was saving energy. Many rap-

Red-tailed hawk sitting out the blizzard.

tors had already died that winter because the snow was so deep it had successfully hidden their prey. He was not about to let me make him battle the raging wind and blinding snow to find another roost.

I drove my car to the end of the driveway, changed my film and returned to shoot more frames from the opposite angle. This time I was six feet from the magnificent bird whose yellow eyes fixed me a cold hard stare. He was damned if he would budge. Even in the low light I could get soft shots of his striking white winter plumage. I shot off another roll and left him to sit out the rest of the storm . . . in peace.

Today I am back at New England Biolabs and find that the same hawk is sitting on a bronze statue of Don Quixote. I'm told the hawk used to sit on the outstretched hands of a statue of Jesus when this property was used as a religious retreat decades ago. Before then, his ancestors probably sat on the cornice of the Proctors' summer cottage: a lavish Victorian Great House surrounded by 126 acres of woods and fields.

But I have returned to investigate a new view of Ipswich and a new view of nature. Soon the Proctors' old home will house the modern con-

Don Quixote.

ference rooms, lecture halls, and administrative offices of New England Biolabs. The lab's weekly lectures promise to draw both scientists and the general public. The Victorian stables, once the home of at least twenty-four hunting and carriage horses, will soon house the Ocean Genome Legacy Foundation, an ambitious project to catalogue and preserve the DNA of every oceanic species, many of whom are already endangered.

A tunnel will connect the main building with two long, glass-enclosed structures that will house labs, offices, and batteries of the stainless-steel fermenters where tiny bits of DNA will be recombined and grown to provide inch-high vials of clear liquid, which will then be shipped by Federal Express to research labs and pharmaceutical labs throughout the world. It will be a clean, high-tech operation that uses minute amounts of raw materials, emits virtually no wastes, and produces a colorless liquid worth several times its weight in gold. It sounds almost too good to be true. How did this come about?

In 1972 Don Comb was an idealistic young biologist at Harvard Medical School. The self-proclaimed hippie was working on the frontier of molecular biology. Before starting an experiment he would have to search the literature for natural endonucleases, enzymatic protein scissors that could identify and snip off tiny discrete pieces of DNA, genes that can determine the color of your eyes or whether you will live to a ripe old age or die at twenty from a crippling genetic disease like cystic fibrosis.

It was clear that this was the wave of the future. Scientists could use these genes to develop vaccines, cure diseases, save and improve lives. But it was a tough slog. Researchers had to find their enzymes in nature. Some came from swamps and marshes, others from geysers and deep sea vents miles below the ocean's surface. Researchers had also discovered ways of amplifying DNA to produce large amounts of these restriction enzymes for research. But it was a long, slow process for an individual lab. What if a company could specialize in mass producing the enzymes? It would save labs countless hours of staff time that could better be spent on research.

Don Comb realized that if he could mass-produce high-quality enzymes and sell them cheaply, scientists would gladly pay for his tiny vials of clear liquid. And there was a large and growing market. Scientists need these enzymes whether they are in universities, private labs, or major pharmaceutical companies. But unlike most biotech companies, the company had an instant cash flow. Don didn't have to worry about selling stocks, marketing drugs, or navigating the treacherous waters of FDA approval. He could keep the business small, private, and in the family. New England Biolabs took off like a rocket and continues to thrive. The company went from producing ten to close to 300 specific restriction enzymes.

Today New England Biolabs even sells their vials in refrigerated vending machines. If a scientist is working on the weekend or late at night he can just swipe a special New England Biolabs credit card through the machine and withdraw enzymes to help him find genes to cure cystic fibrosis or identify Huntington's disease carriers. He could even use one of these enzymes to help create a new strain of weapons-grade anthrax or the vaccine to prevent it. Such are the profound possibilities and potential dangers of our brave new world of biotechnology.

New England Biolabs has become the Ben and Jerry's of the biotech world. Visiting its present headquarters is like entering a cross between an art museum and a plush new university. Young and old scientists walk down the halls, heads bowed in conversation. Others sit in sunny nooks overarched with towering tropical trees or oversee fermenters where the enzymes are grown in batches of the bacterium *E. coli*. Like in a barn, you are never far from the geosmin odor of the fermenters. It is the heady smell of life itself.

The company still reflects Don Comb's humanism. Employees own a third of the company, share in the profits, and have use of a company-owned a villa in St. Barth. Most biotech companies teem with young workers eager to move on to the next higher paying job; New England Biolabs attracts researchers happy to thrive in its unique culture for their entire scientific careers. Dr. Comb has also established the New England Biolabs Foundation, which supports grass roots environmental programs throughout the world. His newest passion is the Ocean Genome Legacy Foundation, a project established to preserve the DNA of key transitional species in the evolution of marine life. His thinking is that most of evolution occurred in the oceans and if the oceans go the way of land, it will be essential to have a bank where the DNA of extinct species can be preserved for replication.

Recently a single sample of water taken in the Sargasso Sea south of Bermuda yielded as many as 53,000 new species and 1.2 million new genes, including 800 new genes for rhodopsin the light receptor pigment that underlies vision. Before this sample was analyzed we only knew of 200 rhodopsin-like photoreceptors. This suggests that the biodiversity of the oceans is far larger than we ever expected and that marine microbes use sunlight in ways we can only imagine. Because of this one sample, scientists now think that more than ninety-nine percent of the earth's microbial species have yet to be discovered, and the sample was taken in an area known as "the desert of the ocean" because it was always thought to be so barren.

This same shotgun method of sequencing of DNA could even be used to sample Martian waters for new forms of life. No wonder Don Comb has a statue of Don Quixote in his driveway. He has a lot of windmills to tilt at. His efforts will be grand but not in vain. We are in the process of rediscovering and reinventing Darwin's magnificent new view of life on this and other planets.

Part V

CONCLUSIONS

PRECEDING PAGE: *The vernal pool.*

Chapter 18

The Big Night

(March 31, 2004)

It is a cold wet night in late March. Raindrops patter quietly on the forest floor and the pungent odor of skunk cabbage fills the evening air. The beams of twenty flashlights skitter nervously from side to side. I have joined a group of Ipswich students to witness one of nature's great events, the annual migration of amphibians toward their nuptial pools. We wend our way single file past patches of ice that still linger along the sides of the rutted road. We trudge on in as much silence as any self-respecting class of fifth graders can possibly be expected to muster on a late-night field trip. It is only their teacher and I who whisper.

Soon even our conversation ceases. We start to hear the duck-like croaks of wood frogs and the high pitched tinklings of spring peepers. Beside us, unseen in the leaf litter, thousands of frogs, toads, and salamanders are crawling toward the shallow pool that lies in darkness ahead of us.

This migration can send a shiver down the spine of even the most obstreperous fifth grader, for it was on a night like this, 365 million years ago, that one of our mutual ancestors crawled out of the mud to escape the shark-like predators that terrorized the rivers and estuaries of a swampy Devonian coast. The ancient tetrapod had already evolved fin-like limbs to clamber through underwater vegetation; all it took was a little evolution to turn the nascent limbs into the walking legs of today's amphibians. The tetrapods had emerged into a food-filled land of fern and lycopod forests, an amphibian Eden, if your tastes ran toward dragonflies and cockroaches the size of small rats.

Tonight the descendents of those amphibian pioneers are returning to these shallow waters to reenact a ritual that has persisted for 365 million years. They seek out these vernal pools because the pools dry up in summer, so no fish can survive to prey on the amphibians' eggs and larvae.

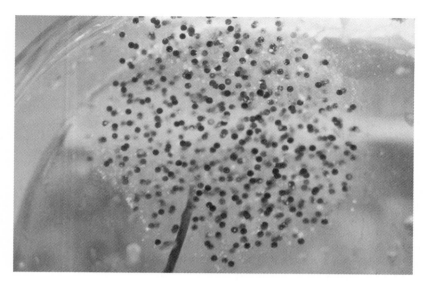

Frog eggs.

The surrounding forest still retains the moss and fernlike descendents of that earlier era, but the pool is now overgrown with modern deciduous trees. The green and purple spathe of a skunk cabbage juts through a patch of snow. It exudes its fetid odor and is surrounded by a ring of melted snow. Both the plant's odor and its internal heat are part of an elaborate sexual deception. The odor is reminiscent of a dead animal, and the heat melts the snow so gnatlike insects think they have discovered a tasty corpse rotting beneath the snow when in fact they have just been cajoled into pollinating the skunk cabbage's ill-smelling flowers.

But now the night air is filled with the sounds of budding young scientists.

"Water temperature?"

"Six degrees Celsius."

"Ground temperature?"

"Seven degrees Celsius."

"Depth?"

"Twenty-three centimeters."

"Do you see any frogs?"

"No."

"Well, put that down too. Remember, zeros are as important as positive data. So how does this fit our hypothesis about the temperature?"

Last night these wood frogs were frozen solid, their blood protected

Frozen wood frog.

by a sugary glucose antifreeze that shielded them from freezing but made them severely diabetic. But yesterday the temperature had risen and the frogs' hearts had started to flush the glucose from their bodies. An hour later they had been mating. They would continue this cycle, freezing then mating, freezing then mating — a remarkable adaptation that allows the frogs to lay their eggs before their predators arrive and before the vernal ponds have fully melted. But the students are trying to discover the exact temperature at which this transition occurs.

Now several students start to wade through the shallow waters. Others stand gingerly on tiny islands of sedge and tussock grass. Sphagnum moss squishes beneath our feet and the quiet calls of a woodcock reverberate through the forest. Suddenly the merest wisp of a tiny creature glides through the beam of my flashlight, then darts beneath some bottom vegetation. Feathery appendages propel the creature that swims upside-down through the water column. Her tiny black eyes stare up at me, hoping I will not scoop her out of the water for closer examination. She had a tiny cluster of eggs protected behind her legs.

This is a fairy shrimp, an evanescent little creature that is far more hardy than she appears. Her eggs can dry up and survive in baked mud for up to twenty years. After a single rain shower, millions of them can hatch out in ponds that have been devoid of the creatures for decades.

Another student finds the inch-long larva of a beetle sequestered in

A spotted salamander.

the squishy soft fibers of a rotting log. Further on, two more students have found a seven-inch spotted salamander making her way purposefully toward the pond. She is fat and heavy with eggs and has shiny black skin covered with large, strikingly beautiful, yellow dots. Tonight the rain and rising temperature are coinciding to advance her pondward migration. Tomorrow night the temperature might drop and she will stay where she is. But she only has two weeks to lay her eggs, so timing and temperature are critical.

Sometimes their window of opportunity is so narrow that all the salamanders end up congregating in a big seething ball of slitherish amphibian delight. Black and yellow salamanders weave in and out of the writhing mass but the ball stays curiously stationary suspended in the water column. It is a little like lathering yourself up with soap and taking a group shower with the lights out and the only sound the quiet squishing of skin against skin. I remember a rumor in college that similar things were taking place in a neighboring dorm, but I never could find out which one.

In any case, the *liebesspiel* has the desired effect. One by one, individual couples break free to perform their own private dance on the vegetative floor. Eventually a male will lead a female to a garden of spermatophores that he has carefully laid on the bottom vegetation. There, the female will straddle his sperm packets and take them up into her cloaca, where the sperm will fertilize her eggs internally.

Other students, sixty-three in all, are cruising the rural streets of Ipswich looking for evidence of amphibian crossings. When a front-seat spotter sights a frog, he signals his mother to stop the car and four kids jump out, flashlights in hand, to count dead and dying frogs, and help the living ones to cross the road in safety. Soon the students are joined by other carloads of youngsters and more flashlights scan the rain-soaked roads. Parents trade notes and teachers snap digital photos in the driving rain. Cell phones crackle as scout cars report to "Salamander One." Self-appointed street guards signal oncoming cars to slow down, only to discover that most are simply more "frog cars." It seems like we are the only cars on the road tonight, and it is probably fortunate. All the flashlights make it look like Halloween, a terrible accident, or that scene in *ET* where the government agents are looking for the lost and ailing extraterrestrial. But tonight these Ipswich students rule the roads. They have convinced the town to put up warning signs at all the major amphibian crossings.

As the reports keep coming in, the students start to see a pattern emerge. Last year the frogs were hopping all over the streets. This year everything seems to be happening in slow motion. Most of the amphibians appear to be stuck, as if they had been paralyzed in mid-motion as they crossed the road. The students gradually realize what has happened. Last week warm rains initiated the migration. Then it had cooled, freezing the amphibians in place. Today the temperature had

Amphibian crossing.

inched above forty degrees for most of the afternoon, then dipped below it again in the evening. When I had left the house to join the group at dusk, the spring peepers across the street from us were calling, but when I return tonight they will be silent from the cold.

We were witnessing the results of crossing this invisible threshold. The amphibians had set out when the temperature had gone above forty degrees. Then, as the temperature dipped below forty degrees, glucose started to build up in their bloodstreams and they once again returned to their hibernation-like torpor. Their hearts had slowed and their muscles had become frozen, leaving them paralyzed, often on the slightly colder macadam. It had been painful, but the students had learned something from the minor tragedy. They had verified that forty degrees was the exact temperature that determines whether the migration will proceed or stall.

This field trip, nicknamed "The Big Night," was the culmination of the students' entire year spent studying vernal pools. Throughout the year they had learned about ecology, making hypotheses and using the scientific method. They had learned how to work together to design an experiment, collect data, and isolate the critical variable that had determined the success or failure of "The Big Night" amphibian migration. It was obvious the kids had been moved and learned something from the experience. Some of these students may one day end up at New England Biolabs using the same techniques to splice genes or grow a new endonuclease; others might end up at similar labs springing up all over the North Shore.

When I returned home that night I decided to make one last visit to the vernal pool opposite our house. I remembered when I had first discovered the pond last spring, when it had been covered with the delicate blossoms of tiny native blue irises. I re-

Pavillion Beach by moonlight.

membered it from last summer, when the pool was entirely dry and I found a doe and fawn grazing among its remaining tussocks of grass and sedge. I remembered it from last winter, when my daughter discovered that the pond made a perfect rink for an impromptu game of hockey. Now I will remember the pool as the first place I became enamored with the spring-time migration of amphibians, the place that plays a critical role in preserving the biodiversity of the surrounding countryside and protecting the life cycles of several endangered species. I will remember it as a place where kids can come to kindle their imaginations and change their lives forever.

Chapter 19

Local Heroes

(April 11, 2004)

The sirens started as we were washing up after Easter dinner. First one, then another hurtling down Jeffreys Neck. Perhaps it was a house fire or medical emergency. But the vehicles kept coming—fire trucks, cop cars, then rescue boats and divers. It was obviously a water accident and it looked bad. Had a boat capsized and thrown several people into the water still hovering at a quickly killing thirty-eight degrees?

I decided to investigate. Evidently other people had the same idea. When I arrived at Pavilion Beach the parking lot was full and fifty people stood staring at the waves. A group of kayakers hugged the beach and more blue lights flashed on the far shore.

Gradually we were able to piece together what had happened. Stephen Thompson was walking around Little Neck when he noticed an aluminum boat spinning wildly out of control. As he watched, an errant wave pitched the operator out of the boat. The boater started to scream for his life as soon as he hit the frigid waters.

Stephen yelled for a neighbor to call the Coast Guard; then he and his father pulled a dinghy, paddle, and life preserver from someone's backyard. With police, firefighters, divers, and the Coast Guard on the way, Stephen launched the dinghy and started paddling toward the swimmer, whose head was now bobbing in the frothing whitecaps. Stephen's biggest fear was that the swimmer would get hacked up by the propeller when the boat made another pass.

Paddling on his knees, Stephen was already getting tired by the time he reached the frightened man.

"Here, put this life preserver on and grab onto the side of the boat."

"Boy, am I glad to see you."

"What's your name?"

"John Barton."

"Well, John, everything's going to be all right. What happened out here?"

"Wave hit me, pitched me overboard."

"Yeah we saw you."

"I'm getting cold, man."

"OK, OK. But don't try to get in the boat. Just keep talking to me so I know you're all right."

"OK."

"Where'd you go to school, John?"

"Ipswich."

"Really? So did I."

"I'm getting really cold, man."

"I know, John, I know."

"What about you man? You gonna make it?"

"We're both gonna make it, John. Now see if you can't kick. It'll keep you warm and it might get us there a little faster."

"You look exhausted, man."

"I know, I know. Hang on, we're almost there."

The next thing the two men knew, a paramedic was rushing out to meet them. She grabbed John under the arms, trussed him up into a gurney, and loaded him into an ambulance.

Stephen slumped in the dinghy, gulping great lungfuls of air. What had just happened? The medic returned.

"Is he gonna be alright?"

"You bet he is, and by the way, Stephen Thompson, you're a hero. You saved his life out there."

"Yeah?"

"Yeah!"

But then Sergeant Moriarity came by.

"You know you're in a whole heap o' trouble, young man. We have thirty witnesses saw you trespass in someone's yard, steal a dinghy and paddle away. What do you have to say for yourself?"

"Can you help me put 'em back?"

"Sure, and I guess we can forget the charges for just this once. But we'll be watching you, Stephen Thompson. We'll be watching you!"

Crane's Beach.

Offshore the boat continued to speed in circles. The Coast Guard had wisely decided to simply keep an eye on the boat until it ran out of gas. Other than the fact that John Barton had taken a cold swim in still frigid waters it had been an exemplary rescue. Everyone would remember this Easter as the day Stephen Thompson saved John Barton.

A few days later a group of volunteers set out on another rescue mission. But this mission would be to rescue a river, not a man. They were already there when I arrived. I joined them beside the swollen Ipswich River, watching its raging waters cascading over the Sylvania dam. Spring floods had already destroyed a section of the retaining wall protecting several shops. One of the shops was already being undermined.

A mother and her daughter peered intently at a white slab of plastic a foot below the raging torrent. Tannin-rich waters flowed over the slab but they could see no fish. Below them a few herring were milling about the dam, searching for the concrete fish ladder that would allow them to scale the dam and continue upstream.

Herring.

    Herring, more properly called alewives or *Alosa pseudoharengus,* are anadromous fish that spend most of their lives in salt water but return back to their native freshwater streams to lay their eggs. Behaviorists think alewives were once freshwater species. But during deglaciation many alewives were swept out to sea. Some of them were able to survive and benefit from the greater amounts of food found offshore; a few of these developed adaptations that allowed them to return to the freshwater ponds to lay their eggs. Over time they developed what scientists call a stable evolutionary strategy, a lifestyle that allowed them to benefit from both the greater amount of food found in the ocean and the greater amount of protection for their eggs provided by freshwater ponds.

    This unique behavior makes herring the perfect indicator species to monitor the health of rivers. Every spring, alewives must ascend narrow fish ladders right below the feet of eager human fish counters. The other trait that makes herring an ideal indicator species is that they return to the same stream in which they were born. But how do they know which

Eagle Hill River.

stream? Scientists discovered how anadromous fish discern their natal stream by devising a unique and slightly whimsical experiment. First they put red balloons on all the fish that had swum up the right fork of a river and put blue balloons on all the fish that had swum up the left fork. When they took the fish below the fork and released them, all the balloons were mixed together until they reached the fork, and sure enough all the red balloons swam to the right and all the blue balloons swam to the left. But then they dug up some sediment from the right fork and placed it in the left fork. This time, when all the balloons got to the fork they milled around for a minute, then swam directly up the left stream. The fish had honed in on the unique smell of the sediments in their own native stream. They had imprinted on this unique odor as tiny inch-long fingerlings several years before.

But the mother and her daughter didn't expect to get such dramatic results. In fact, they might not even get to see a single fish. For the last five years, volunteers have only counted 200 herring. In historic times the Ipswich River teemed with large runs of herring, smelt, shad, and salmon. The nearby Parker River had 38,000 herring in 1974; now they are down to less than 2,000. Why? The reasons are many. Dozens of cities and towns draw so much water out of the Ipswich River that it runs dry almost every summer. Every time the river does run dry, it kills off many of the river species that need colder, fast-running water, but the pond species survive. Perhaps these species that live in the warmer water ponds where alewives lay their eggs consume more of the larval herring than the river species would.

Certainly this dam doesn't help. It was built in the 1880s to power the mills that produced Ipswich lace. Later it was used by the Sylvania plant that made light bulbs. There are some plans afoot to remove the dam, but some environmentalists urge caution. Two hundred and twenty-five years worth of sediment sits behind that dam. Nobody knows how many heavy metals might reside there. What could happen to Ipswich's famed shellfish industry if you released all these sediments at once?

Foreign fishing fleets.

Some scientists even point to offshore fishing as the culprit. During the 1970s foreign fleets fished down most of the cod, haddock, and flounder that prey on herring. Herring started to briefly rebound after the 200-mile fisheries limit went into effect in 1976.

So what were the mother and her daughter doing out there counting fish in what has been declared the third most endangered river in the nation? Perhaps their work will help convince the state to restock the river; perhaps their counts will help convince towns and cities to stop drawing down so much water to flush their toilets and water their lawns.

But perhaps the greatest impact is that every day volunteers are now out there collecting this data. Previously a graduate student might have collected these numbers and used them to publish a paper whose primary purpose was to help him ascend his own ladder of professional success. Today, the Ipswich River Watershed Association has seventy-five volunteers who count herring and monitor ponds. They and their families have a chance to become intimately involved with the river and become advocates for its improvement. Maybe they should be trying to

Time to rest.

stop global warming; maybe they should be working to stop wars in the Middle East; maybe they should be working to stop the kind of global thinking that leads to such wars. But maybe, on a beautiful day in early spring, it is just quite enough to be trying to save this quiet corner on the North Shore of Boston.

Index